Raising Chickens

Beginners Guide to Raising Healthy and Happy Backyard Chickens

By: Janet Wilson

ISBN: 978-1-951791-59-9

Table of Contents

Part One: Introduction 1

Why is it worth considering whether chickens are for you?........................... 2

Frequently Asked Questions 3

    How much time and work will it take to keep backyard chickens? 3

    Are chickens expensive? 5

    Do chickens smell? 5

    How do I know how many chickens I can have? Can I have one chicken? 5

    How many eggs will I get?..6

    Can people get sick from backyard chickens?..6

    Am I allowed to have backyard chickens? City Ordinances and Homeowner Associations 8

    What's a wattle? 9

    The Crop 14

    The Cloaca and Vent 14

Part Two: Getting Started 15

Decide how many chickens you will have.. 16

Contact a Local Chicken Expert if Possible 17

Bare Basic Requirements for any Backyard Chicken Coop............................ 17

Choosing or Building a Coop.. 18

    Chicken Tractors 19

    Chicken Coops...22

    Building Your Chicken Coop 28

    Chicken Coop Accessories 39

Bedding **45**

    What are the functions of bedding in your coop? 46

    In the Outdoor Run/Enclosure 49

    A Bedding Accessory – Diatomaceous Earth (DE) 50

    Coop Bedding Conclusion 52

Choosing Your Chicken Breeds **54**

    Australorp 57

    Rhode Island Red 58

    Plymouth Rock (aka Barred Rock) 59

    Speckled Sussex 60

    Wyandottes..62

    Leghorns..64

    Dominique (aka Pilgrim Fowl) 66

    Buff Orpington 67

    Jersey Giant..68

    Isa Browns 69

    Welsummers 70

    Salmon Faverolles 71

    Easter Eggers..73

    Brahmas 76

    Polish..78

"Bantam" Breeds **80**

    Silkies 80

Mille Fleur d'uccle (aka "Mille Fleurs") 82

Frizzles..83

Online Hatcheries **84**

Roosters **85**

Brahma Roosters...86

Orpington Roosters...87

Faverolle Roosters..87

Plymouth Rock Roosters 87

Black Australorp Roosters 88

Welsummer Roosters...88

Cochin Roosters 89

Chicken Feed and Nutrition.. **91**

Basic Nutrition Needs 91

Water 91

Chicken Feed..92

Which feed grind should I give to my adult chickens? Pellets, mash, or crumble? ..93

Feed Storage..94

Grit **94**

How to Give Your Chickens Grit 96

Grit for Chicks up to 8 Weeks...97

Grit for Chicks 8-18 Weeks (aka pullets) 97

Grit for Adults 18+ weeks..98

Can free-range chickens get enough grit from just pecking around outside? 98

Storing your grit 98

Treats **99**

    Meal Worms 99

    Treat Mixes 99

    Seeds 101

    Feeding scraps..102

    There are a few food scraps to avoid:..102

Cleaning the Coop **103**

Chicken Manure and Compost **104**

Collecting Eggs **104**

    Cleaning your Eggs 106

    Storing your Eggs 108

Trimming Claws, Spurs, and Beaks **109**

    How often will I need to trim claws, spurs, or beaks? 109

    Trimming Claws..109

    Trimming Rooster Spurs..111

    Trimming Beaks..112

A Well-Ordered Pecking Order **113**

Introducing New Hens to Your Flock **114**

    Set up for Success 115

    Move your Chickens into the Flock 116

    The First Week 117

Caring for Chickens in Cold and Hot Conditions **117**

    Chickens in Cold Conditions 117

    Chickens in Hot Conditions 121

Caring for your Chickens in Hot Conditions 122

Conclusion of Caring for Chickens in Hot and Cold Conditions **123**

Molting **124**

How to Care for Your Chickens During A Molt 126

Part Four: Chicken Dinner **128**

Killing a Chicken **129**

Killing a Chicken with a Pellet Pistol 129

Preparing your Chicken 130

Part Five: Breeding Chickens **132**

Breeding Behavior **133**

Hen Care During Egg Development 133

Congratulations! Hatched Chicks! Chicks up to 8 Weeks 134

Introducing Pullets or Young Adults to the Flock 135

Using an Incubator and a Brooder 136

Breeding "Mutts" **138**

Part 6: Troubleshooting **140**

Common health problems **141**

Diseases and Infections **141**

Fowl Pox 142

Infectious Bronchitis 142

Newcastle Disease 143

Coccidiosis 143

Bumblefoot 143

Botulism 144

Thrush 144

Infectious Coryza 145

Fowl Cholera 145

Pullorum 145

Parasites **146**

Red Mites (Dermanyssus gallinae) 146

Scaly Leg Mite 147

Mites of the Feather Shafts and Bases 148

Fleas 148

Ticks 148

Worms 149

Wounds **150**

Stop the Bleeding and Assess the Wound 150

Treat the Wound 150

Isolate the Chicken While the Wound Heals 151

Sour Crop **151**

Egg Bound Hens **152**

Broodiness **152**

How to Tell if a Hen Has Gone Broody 152

How to Prevent or Discourage Broodiness 153

What to Do if You Have a Broody Hen 154

Soft-Shelled Eggs **156**

The Pecking Order vs. Bullying and Cannibalism **157**

The Pecking Order 157

Bullying and Cannibalism 158

Rooster Trouble 160

A Word About Dogs **165**

Conclusion **166**

Footnotes **168**

Part One: Introduction

Backyard chickens have taken off in popularity. Many people in suburban and some urban areas are raising chickens, and the numbers are on the rise – in 2018, over 10 million households had backyard chickens.[1] You can purchase or build a coop that is limited only by your needs and imagination and fill it with the right number of chickens for your space.

Why is it worth considering whether chickens are for you?

1) They give you food. *Really. Good. Food.*

Have you ever had an egg for breakfast that was laid that morning? They are shockingly different from store-bought eggs. You may already know the difference in quality between free-range pastured chickens and those that were factory-farmed, but having your own chicken just butchered, takes the quality to another level.

The best eggs you've ever tasted? Worth all the effort.
Your child coming inside, wide-eyed, with an egg they can eat for breakfast? Priceless.

2) Chickens are great gardening partners![2]

When your vegetable garden is complete for the year, let your chickens out on the space, and they will clean up any weeds, old roots, and any leftovers of vegetables. They will also aerate the soil with their scratching and pecking, fertilizing the garden bed as they go. If you have a portable enclosure (called a "chicken tractor"), you can fertilize and aerate an entire flat landscape.

3) Chickens keep down flies, beetles, and other insects. I've had personal experience with this.
   - The first year that we moved onto our land, we were amazed (and a little horrified) at the number of beetles, moths, flies, and other insects. The number of pine moths was absolutely out of control.
   - Year two, we introduced six chickens, and we were amazed at the difference: less than ½ of the pine moths and other insects.
   - Year three, we did not have chickens because a marten, fox, and dog managed to kill the chickens. That summer, the flies, moths, and beetles were up again (not as much as the first year, but certainly more than the second year.

- <u>Year four,</u> we got our protection sorted out and introduced more chickens (and a flock of baby chicks). The insects are much lower than year two.

4) If you have kids, they will learn to care for animals and know where their food originates.

   There is a natural joy for a child when they can take part in caring for and gathering the eggs from the chickens. There are also the realities of a little hard work – but small enough to be manageable (unlike larger livestock). They will learn a lot, and the entire experience can be a way to strengthen the bond of the family.

5) Chickens are entertaining, relational, and beautiful.

We will talk more about the details of breeds that are gentler and more interactive. What we can say here is that generally, they are relaxing and enjoyable to watch. One author called it a "zen experience" watching "chicken television" with his kids.[3] They have personalities and dynamics. You can watch them like an ongoing TV series.

6) Chicken manure will compost well and is an excellent organic fertilizer.

There are a lot of common questions that people have when considering whether backyard chickens are the right choice for them. We've collected the most relevant FAQ's to help you decide.

Frequently Asked Questions

How much time and work will it take to keep backyard chickens?

There is a fair amount of work to do when you set up. You are researching, learning, making a lot of decisions about placement, breeds, the number of chickens to have, getting the coop and the feed set up.

After the setup, you've got the chickens and can go into maintenance mode.

I've found out that having six chickens in my backyard took less time and care than caring for my dog.

Here's a sample schedule for maintaining your chickens once you have set them up:

- <u>Every morning</u>: Let your chickens out of the coop and into their enclosed run. Change their water, surface clean their coop, and check for eggs.
  - Note: Some people choose only to clean the coop once a week. That makes it a bigger job, but not daily. Similar to dog poop, the longer you leave the poop, the more there is to get on the eggs and cause bacterial issues. My preference is to clean out the poop from the nesting boxes as well as the surface of the coop and enclosure daily. Then do a change of bedding and thoroughly clean the cage every week. In the mite season, I thoroughly scrub the coop weekly.

Another coop cleaning method is called the "deep litter method." Deep litter minimizes cleaning the coop to every six months. The idea is that instead of removing the chicken manure regularly and composting it separately, you let it compost in the coop and then clean it out and start over. 4-6" of dry wood shavings lasts six months. Instead of cleaning the coop, all you have to do is add more bedding along with some treats to encourage the chickens to peck, scratch, and aerate the compost. The compost from this process is fantastic, and it can go straight to your garden beds. There are pros and cons; we will discuss the details of this later.

- <u>Afternoon</u>: Check for eggs again (if a hen is laying, she is likely to lay twice a day).

- <u>Late Afternoon/Early Evening before dark</u>: Lock them in their roost.

- <u>Once a week</u>: Clean the coop if you're not doing it every day[4]. If you can allow your chickens out to free-range instead of being held in an enclosure, then that's a good one. Some people say they only need to clean the coop every 3-4 weeks. [5] Do your research, and if possible, get advice from a local chicken expert about your specific backyard situation. Personally, in an enclosed space, I'm all for a bit of work every day rather than a lot of work once a week, and I also want to have clean eggs.

- <u>Other random extras</u>: Like any domestic animal, you'll need to pay attention to any signs of health concerns (will be discussed later). You may also want to give them treats in the afternoon. Chickens need their feet cleaned when you handle them – this can be a super quick wipe off once you get the motions down, and it's important to make them more comfortable handling them properly and checking their feet.

Are chickens expensive?

Setting up is a bit of outlay: $100-$700 (or more if you choose) for the coop, fencing, and set up. That's a wide range because it depends on whether you build it (later, we will feature a video on how to build a coop for $50 in an hour) or purchase a pre-made coop, which can run even higher than $700.

There is a wide range of setup cost: you don't have to spend a lot of money. After that, $20/bag for 50 lbs of feed. One bag feeds a flock of six for a month.

You won't have a lot of vet bills if you follow the advice in this guide concerning space, keeping the coop and enclosure clean, as well as being watchful for any wounds or illness in your flock. Giving them plenty of space is also key to their health.

Generally, people who have chickens say that they quickly recover their cost in eggs, those who also eat the meat even more so.

Do chickens smell?

Chickens themselves don't smell, but their poop does. You'll have a strong, unpleasant smell if you don't change that coop and refresh it with new bedding at least once a week.

Don't let chicken manure alone put you off of having chickens. We will discuss how to compost your manure so it doesn't smell or attract flies, but rather, it provides you with fantastic fertilizer.

How do I know how many chickens I can have? Can I have one chicken?

Chickens are flock animals. You need more than one.

The first thing you need to do is figure out how large your coop and enclosure will be. When you know the square feet of the total area, you can calculate how many chickens to get. This guide describes the details of this process in Part Two: Getting Started.

How many eggs will I get?

There isn't an easy, straightforward answer to this question because breeds, strains, age, and individuals vary. Also, the conditions make a difference. Another factor is that they do not lay eggs evenly through the year; with few exceptions, they mostly stop when there are less than 12-14 hours of daylight. They also slow down substantially when they are molting [6]

Ensuring excellent nutrition will also increase egg production. You can see that there are a lot of unknowns at play, trying to calculate the number of eggs that you will get.

One way of calculating is by the number of family members. You could start with two to three chickens for every two family members. If you want to make sure you have plenty and sometimes some eggs to give away, then make that two to three chickens per person.[7]

There are chicken breeds that are "layers" and others that are raised more for their meat. (We will discuss this in more detail when we discuss breeds.) The layers will typically lay 200-300 eggs a year; remember that is not divided evenly throughout the year.

Can people get sick from backyard chickens?

Yes, you can get sick. The CDC raised the alarm about salmonella cases in 2017 and again in 2019. In August 2020, there was yet another article about increasing salmonella cases that were traced to backyard chickens. [8] Most of these were *not* related to the eggs but from the *handling* of the chickens.

The CDC supports the trend of backyard chickens. They provide guidelines and simple, common-sense protocols you can follow to avoid getting sick and alleviate the concern.

The three major points are:
- Wash your hands,
- use outside-only shoes,
- Don't let your chickens inside.

Here's more detail of the CDC's guidelines:[9]

First, you must know that a chicken with salmonella can look completely healthy. You must follow these protocols even if your flock is a beautiful specimen of health.

Wash your hands thoroughly with soap and water after collecting eggs or handling the chickens or their coop/roaming area. You can hold them, have them on your lap and be very relational with them, but you must wash your hands afterward. Use hand sanitizer if soap and water are not available. Make sure that children do that as well, and endeavor to supervise them to ensure they get the job done right.

Do not allow your chickens or baby chicks to come into the house. (They walk around in their own poop and have bacteria on their feet and feathers. They may also peck something in your house and leave bacteria on the item.)

Have a pair of outdoor only shoes for each member of the family by the back door. You don't want to track in the bacteria.

Children under five must be supervised by an adult when they are in the chicken area; they might put something in their mouth and are likely to touch their mouth or eyes after picking up or touching a chicken or something in the yard.

You can hold your chickens, but don't cuddle them next to your face or kiss them. (Yes, this is a problem – especially with kids). Chickens peck in around their poop. Their beaks, feet, and feathers can have bacterial contamination.

Clean the feeders, water containers, and any other equipment outdoors. DO NOT do this in your kitchen sink.

Collect your eggs twice a day. The longer they are out there, the more likely they will be contaminated. Inspect your eggs for cracks and throw away those that even have hairline fractures.
If your eggs have dirt or debris, you can use a designated piece of fine sandpaper or a brush to get it off. T*he CDC notes that you should not wash the eggs because "colder water can pull bacteria into the egg."* [10] The open pores of the egg will *pull the bacteria in* from the outside. We will discuss more details in the egg section later.

Of course, cook your eggs thoroughly.

By following simple protocols, raising a few chickens would not need to be a health concern.

Am I allowed to have backyard chickens? City Ordinances and Homeowner Associations

Chickens used to be encouraged by the federal government. This poster is from 1918:

One hundred years later, things are different. It surprises many, though, that many large cities allow backyard chickens. Chickens are permitted even in New York City. [12]

The first thing to do is to check the ordinances in your municipality. The details of the regulations will vary, so make sure you know the rules even if you know that they are allowed. Most laws are concerned about them being a nuisance of noise or smell, so they have regulations for how many chickens you can have, and typically, roosters are not allowed.

If your city does not allow them, you can advocate overturning the ban. There are increasing numbers of victories, including in Madison, Wisconsin; Ann Arbor, Michigan; Longmont, Colorado and Bozeman, Montana.

Please check out this footnote for some great resources for chicken advocates. The first link is for a handy .pdf that is an advocacy guide.[13]

Once you've confirmed that your city allows chickens, then the only barrier you may have is if you are part of a Homeowner's Association. Check your HOA rules. If chickens are not allowed, you can use the resources above to start a petition and advocate for a change to the bylaw that disallows them.

What's a wattle?
Let's go over some fundamental chicken anatomy.

Combs, Wattles, and Lobes

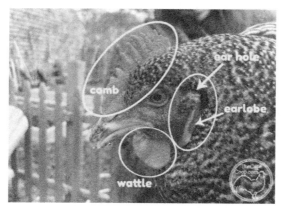

"The Cape Coop" http://ow.ly/ClTt50CszFm

Combs, wattles and lobes are the non-feathered red fleshy bits around a chicken's face. Chickens don't sweat, so their heat regulation is through their wattles, combs, and lobes. Usually, chickens with large wattles and combs are better suited to hot climates. In contrast, chickens with smaller wattles and combs are better suited for cold temperatures to retain heat and reduce the risk of frostbite.

The comb is on the top of the head. The shape of the comb varies by breed. Here are the forms you will see. Sometimes when the feathers of breeds are similar, the combs will be the best identifier of that breed.

Let's look at the two most common combs and the comb on the Brahma chicken, which is featured later in breeds.

Single Comb
The "single comb" featured in the photo below is very common. Many of the breeds we feature in this guide have single combs. These combs can vary widely in size, from big and floppy (like those on the Leghorn chicken) to the small and compact (like those on the Orpingtons).

The Happy Chicken Coop http://ow.ly/cUZ650CsACU

Rose Comb

A rose comb is on the top of the head, and it's flat. Below is a photo of a Dominique chicken.

Nantahala Farm http://ow.ly/SjPV50CsASL

Pea Comb

In contrast, this is a tiny comb, which is directly above the beak instead of at the top of the head. This photo is a Brahma chicken, one of the breeds that we will be discussing later.

Self Sufficient Living http://ow.ly/go8950CsBcY

Earlobes

The most interesting thing about earlobes is that they may tell you the color of egg a hen lays. A chicken with white ear lobes will lay white eggs, while a chicken with red earlobes usually indicates brown eggs. The fun begins with the "Easter Eggers." You can tell whether the eggs will be blue or green.

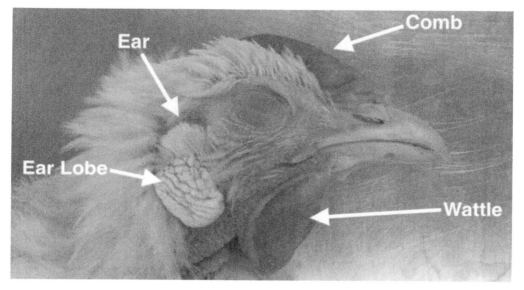

McGill.ca http://ow.ly/ZEbz50CsBuf

Below is a diagram of the anatomy of a chicken. We are only going to look at two more parts of the chicken, as they are essential to understand and will later come up in this guide.

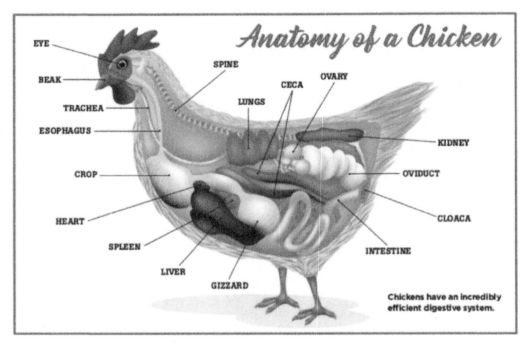

Hobby Farms http://ow.ly/tQ7x50CsBQp

The Crop

Notice the white part under the throat of the chicken labeled "crop". The crop is a pouch where food is stored and broken down before going to the stomach. Chickens don't have teeth, so this is how they "chew."

The Cloaca and Vent

Chickens lay eggs, poop, and mate from the same orifice, the cloaca. The vent is the opening of the cloaca.

Part Two: Getting Started

Decide how many chickens you will have.

It is easy to calculate how many chickens you can have.

First, you will need to measure the total space you have got for a coop and enclosure and calculate the square feet.

Here is the most straightforward "thumbnail" estimate:
- Allow for 4-5 sq. feet of space for each standard-sized chicken in the coop (e.g., that would be 4 chickens in 4X4 (16 sq. feet) or 5 chickens in a 5X4 (20 sq. feet) coop.) *and*
- 10 sq. feet of space per standard-sized chicken inside the enclosure.

These numbers will depend on whether you are choosing large chickens such as Australorps or Brahmas or small chickens such as Bantams. If you want to refine your space estimate, here are more considerations:
- Do you have a small enclosure or during freezing winters, where the chickens will need to be inside for extended periods? If either of these is true, you will need more space in the coop.
- Do you plan to get more chickens? I always like to have fewer chickens than the maximum; make sure there is ample space to introduce new chickens if you want the option to have them.

So, for example, if you want a flock of 6 chickens, you need a minimum of 90 square feet in total. You will need a 6X5 coop and at least 60 square feet in the enclosure *outside* the coop.

It is important to note that there are coops on the market that will claim to hold more chickens, but they may not be healthy for the chickens. By "healthy," I mean that they may not be good for their wellbeing and may also expose the chickens to diseases and destructive behaviors. Measure the space and consider the size of your chickens. Do not just go by a manufacturer's claim.

If you are a beginner, we recommend that you start small. If you have enough space for twelve chickens, that's great, but still start small. Build a coop and enclosure for however many chickens you want. But do not get that many chickens at once, and do not fill it up to maximum capacity quickly. I'd suggest that you start with ½ as many as you can have, get used to having chickens, learn about them, take care of them, get the routine in your family, and learn about adding chickens to your flock. If you can have more than eight chickens, I would still recommend not starting with more than four.

Contact a Local Chicken Expert if Possible

Guides such as this one, as well as books, articles, websites, and forums, are very important and useful to enable you to get solid knowledge for raising your chickens and making decisions about how they will work in your home. There is a lot of reliable information available. People can successfully raise chickens by doing their research, buying or making a coop and enclosure, going to a local hardware store, and purchasing the chicks they want.

If you need to do it on your own, it is totally doable.

Even so, it is worth trying to find a local chicken expert to talk to personally. Nothing can replace the connection with a local expert for asking questions that come up. Also, this person may be the trusted source for the chickens you buy.

If you are going to purchase from a store, ask questions, and find out where the chicks or chickens have been bred and raised. You want to ensure that they are from a trustworthy source so that you are not getting diseased chickens at the outset.

The reading and research you do will give you a solid base of understanding to be able to make the most of your time with an expert. Your self-education, along with a personal guide, is the best possible combination.

Bare Basic Requirements for any Backyard Chicken Coop

No matter the kind of coop you choose, you need these necessary things:

- Nesting Boxes

If you want those eggs, then give your hens a nice private dark space. Tip: make sure there is a door at the back so you can access the eggs easier. 4-5 hens will fit in a box that is 14″ W x 14″ H x 12″ deep.

- Bedding
  We will talk about bedding in a later section. Humans like fresh, clean, soft bedding – so do your hens.

- Perches/Roosting Bars
  A wooden roosting bar off the ground feels right for them. Their instinct is to sleep on a high branch. Perches are sold in both plastic and metal, but chicken feet can find it hard to grip them, so many experienced chicken raisers recommend wood. Each chicken needs 11-12″ of space. Also, note that the roosting place should be the highest point in the coop. If the nesting space is more elevated, they are likely to sleep in their nests, which will be a mess for you.

- Hanging Feeder and Water Source
  We list "hanging" as a necessary point because of the attraction of mice and rats.

- Dropping Tray or Board
  Dropping trays are a life hack for raising chickens! If you have a removable tray or flat board at the bottom of the coop, it will make cleaning every day *much* easier.

- Enclosed Run
  Make sure you have enough space (the calculation of this was discussed earlier) as well as metal fencing that is at least 12 gauge. If you need to keep out aerial predators and clever climbing mammals like raccoons or various weasels, the top must be covered as well.

Choosing or Building a Coop

Chickens care about having an airy, dry, clean space that provides protection from predators, heat, and cold, along with a place to nest and a room to perch for the night. In terms of a coop, that's it. Chickens don't care about anything else. Apart from their needs, the coop you build or select is an individual as you are. The amount of space, your specific needs (such as consideration of children), as well as your aesthetics all come into play so that you can set up your coop to make you as happy as your chickens are.

You may have the space to allow your chickens to be free-range. Don't make the mistake of thinking that you don't need a coop at all. One problem with this is that they will nest all over the place and hide their eggs. There may be weather conditions or the presence of a predator that has discovered them and pose a threat to them, and that will require the need for protection and security. Providing attractive nesting space and protected space to roost is strongly recommended.

Earlier in the FAQ, we mentioned the "chicken tractor" to help you with a large flat landscape. Let's have a look at some examples.

Chicken Tractors

Chicken tractors are a great choice if you want a small flock of chickens and you have a large flat space such as a lawn.

The concept of the chicken tractor is that the coop is on wheels, and the enclosure is attached with handles to enable you to move it. The chickens feed on the ground inside the enclosure for 24-48 hours; then you move it to the next space.

You can purchase or build your chicken tractor. Here is one to buy for $728.00.

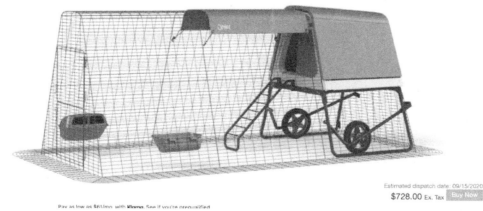

Estimated dispatch date: 09/15/2020
$728.00 Ex. Tax Buy Now

Pay as low as $61/mo. with Klarna. See if you're prequalified

Omlet http://ow.ly/13iQ50CsBWW

19

If you prefer wood, this next one is charming. There are removable roofing poles for easy cleaning and access, two metal trays, and two sleeping ramps. The wood has a waterproof coating. $389.99

Freddie Duplex Chicken Coop on Wayfair http://ow.ly/BFft50CsDvd

Or you can make one such as this for very little money—the designer for this estimated $50 in hardware costs.

The designer gives directions here on backyardchickens.com http://ow.ly/ryZ650CsDDS

If you have a flat space, a chicken tractor can be a great choice. You will still need to clean your coop, but not shovel manure from the enclosure. Instead of shoveling your enclosure every day, all you will need to do is move it around and let the chicken manure fertilize the lawn or selected garden space. Note that pulling it over rocky or bumpy landscapes are not advised, and if the bottom of it is not level to the ground, it makes it easier for predators to get in.

If you decide to build a chicken tractor, you will need to be very conscious of the weight. One blogger described his excitement about making his chicken tractor; it was absolutely adorable and well built. "Well built" in his words like "a battleship." [14] In the end, it ended up being a stationary chicken coop and enclosure instead of a tractor because it was too heavy to move.

The light chicken wire is not recommended for chicken tractors unless it is used along with other higher gauge metal fencing. It won't hold up in the long term with the movement, and predators are well versed in breaching chicken wire. The chicken wire keeps chickens in, not predators out.

From DIY Projects: http://ow.ly/LfsS50CsDJq

This woman offers plans to this chicken tractor in small-medium-large sizes.

Again, note the large flat space around these chicken tractors. If you have a space like this, a chicken tractor might be a beneficial thing for you. The chickens will fertilize your grass, aerate your soil, and gobble up bugs. Also, you don't have to deal with the maintenance of a stationary chicken enclosure.

You may not have the space and landscape to accommodate a chicken tractor, or you may want a larger flock of chickens.

Let's look at standing chicken coops.

Chicken Coops

After you have the measurements for how large your coop and enclosure will be, you need to decide whether to purchase a coop, build yours from purchased materials, or put together a coop with recycled materials.

Chicken Coops to Purchase

There is a wide range of prices and designs for chicken coops to purchase. There is also a wide range of quality. Also, note that chicken coop sellers will often say that their coops will hold more chickens than they would consider healthy. Use the measurements and do the math yourself to determine how many chickens every coop will accommodate.

Here are some curated examples from reputable sellers. The employees of these companies are knowledgeable if you need to ask questions. You may find a local person who would build you a custom chicken coop, but if you get ideas and a sense of materials using these resources, you will be able to negotiate the design and the building of your coop with more confidence.

We have no affiliation with any of the following companies. Still, we include them here because they are examples of reliable, durable, well-thought-out chicken coops with companies that offer customer service.

This one is from Omlet. Besides offering great details and features, they are also designed to keep your chickens warm *and* cool. You can also purchase fitted "extreme temperature blankets" for any coop you choose. They are US based, and there are knowledgeable and experienced individuals available for helpful customer service and advice.

Esti
$

Omlet http://ow.ly/t6mn50CsDW9

If you want something wooden and can pay for top quality, you can purchase a handmade coop with Amish craftsmanship. You can search this site by style or number of chickens. This is an example of a 6-hen coop. $1,650.00

Hand Made by Large Chicken Coops http://ow.ly/LEJ950CsE0W

Pinecraft is another Amish company. Beautifully made and well thought out, if money is not an object, you can find many charming options on their site. Here is one example: $3,276.00.

Handmade by Pinecraft http://ow.ly/BISg50CsEcX

Reading reviews will tell you that you have to be very careful about purchasing a chicken coop on Amazon. We are listing this site because it has strong reviews and a 12-month warranty. A lot of thought went into access and ventilation, as well as aesthetic design. $333.99

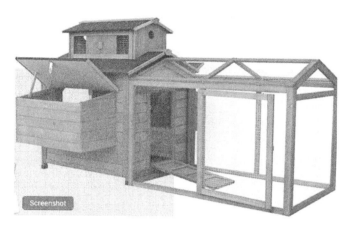

On Amazon offered by Lazy Buddy http://ow.ly/Uhva50CsEqC

Here's another affordable wooden coop that's made of strong fir. At the time of writing this, it was available, and the reviews were good. Check again before you buy it. $294.99

On Amazon offered by Pawhut http://ow.ly/4mcK50CsEBc

Petco offers this small coop with a 1-year warranty. $303.99 It also has strong reviews.

Petco http://ow.ly/Csjf50CsEGg

There are more wooden coop examples out there, be careful when you shop. Many have terrible reviews, and many are listed as "currently unavailable" with no certainty that they will ever come back. A wooden coop is either more expensive or made very cheap with flimsy materials. That is why a lot of people who choose to have a wooden coop decide to build their own. The lower-priced examples above are the best ones we found as quality affordable options.

Snap together chicken coops are another option. These offer ease of cleaning and affordability. One of their significant advantages is that you can get any parasites out, such as the red mites that like to hide in the crevices of the wooden coops. These snap-together coops can also "blend in" if you don't want to pay the price for a wooden coop. Both of these coops would require you to make or purchase your own fenced enclosure. This larger one is $795.00

Large Chicken Coops http://ow.ly/JEvu50CsENS

Rita Marie's also offers a smaller option for $495.

Large Chicken Coops http://ow.ly/JEvu50CsENS

Building Your Chicken Coop

You can build your coop with a plan, re-purpose/recycle something else to become your chicken coop, or, build your coop with recycled materials.

Whatever plan you choose to build your coop, DIY can be an option to keep costs down. It doesn't have to be hard. It can also be an opportunity for a family project leading up to getting your chickens.

In this section, we will look at:
- Building your coop from a plan
- New and used materials for building your coop
- Ideas for re-purposing or recycling something else to be your chicken coop

Building your coop from a plan

Whether or not you have building skills or experience, numerous helpful sites offer free chicken coop plans. They are so numerous that our job in this guide is not to find them but to curate out of the overwhelming amount of resources.

The examples listed here have met the following criteria:
- If it leads you to believe that the plan is "free," then it is. **Not** "oh, the free one is really complicated, and you have to install a 3D program that's hard to understand if you're not a graphic artist – here's a .pdf that you can purchase for $15."
- The instructions are clear and well presented.
- The difficulty level is clear and accurately presented.
- Any downloads are accessible. Is it easy to tell the difference between where to download the .pdf for the plan and where to click on advertising?
- There is evidence of a competent person behind the site, providing payment assistance or limited free access to questions.

Bear in mind that it is recommended that when you set up your main coop, you will need to build a small extra coop in case you need to protect a hen raising some chicks, isolate a pecker, or quarantine a chicken due to illness.

I love this site for free coop plans because they let you pick the number of chickens you will have, a thumbnail price range based on $, $$, and $$$, and the level of difficulty you want. They are clearly labeled "easy," "medium," and "hard."

I searched for 1-6 chickens, lowest price, selected "easy", and I got this plan. They say you can make this for under $150.

The Happy Chicken Coop http://ow.ly/jKgU50CsF44

There are many more free plans, including larger, fancier ones, and purchases are available. Everything is clear and accessible.

buildacoop.com is a reliable and trusted source for plans and great advice about raising chickens generally.
- They have free plans as well as e-books for purchase.
- There are sections for small and medium-sized coops.
- The site expects that the builder will take 24-72 hours to make the coop once the site is prepared, materials are purchased, and tools are laid out and ready. The length of time to build will vary according to the experience and skill of the builder(s).

The site asserts that *"If you can swing a hammer, make a few simple cuts and follow instructions, you'll be able to follow any of our coop plans and build the perfect coop!"*

- There are sample plans for you to look at so you can determine whether this is a good choice for you.
- Each coop plan also has a video with graphics that show the building process. These can complement the instructions that come with the plan.

Here is an example of a medium-size coop from buildacoop:

Build-A-Coop http://ow.ly/ZMCk50CsF82

Home Depot offers a free, straightforward plan for this chicken coop. It is well thought out with ventilation, egg access, and easy cleaning. The instructions and materials listed are very clear, and I would expect that if you had questions, there would be some experts at your local Home Depot who would help you. Notice that you would have to buy or make your own enclosure.

Home Depot Plan http://ow.ly/SwDk50CsFdI

Below is a straightforward plan, and the materials list is right there on the website:

Down East Thunder Farm http://ow.ly/qfVE50CsFhF

We *almost* did not shortlist the plan shown below because of the amount of advertising on this site. We decided to include it because it costs less than $100 to make, and the plans are accessible and straightforward. You will have to put up with a lot of advertising flashing at you, but it may be worth it.

Good Home Design http://ow.ly/VSEF50CsFUc

If you go to this footnote, you will find resources that list other plans for chicken coops. Most of them do not meet the criteria we listed above, but you might find exactly the right one for you if you look around further.[15]

Materials for Building Chicken Coops

The primary criteria for chicken coops are:
- Are the materials sturdy and durable?
- Are they strong enough to provide protection from predators?
- Are they non-toxic?
- Will the material attract pests such as mites?
- What is the cost?

Wood is the most common material used for chicken coops. Hardwoods such as tropical wood, redwood, or cedar (we will talk more about cedar) are more pest-resistant than softwoods. Most coops end up getting built with the more affordable softwoods such as pine or fir and then stained or painted with a non-toxic product. Wood soaks in smell, bacteria, and any liquid and may be prone to leaking.

Pressure Treated Wood

If you use recycled wood, you don't want chemicals going into your eggs and chicken meat. Wood that is older than 2003 has chromate copper arsenate (CCA), which includes arsenic. CCA is very poisonous, and "long term exposure to the arsenic, which is found in some types of CCA-pressure-treated lumber, can increase the risk of lung, bladder, and skin cancer over a person's lifetime."[16]

It's not only long-term exposure that can be a health hazard. "The arsenic levels in the wood have also been shown to have a negative health effect... for example, when children touch wood in play areas, and then put their hands into their mouths."[17]

For more information on the risks of pressured-treated wood, visit this site.[18]

Chicken coops made to spec by a plan with new wood are very attractive, but you can also use recycled wood to reduce the cost. Just heed to the warnings above on the wood you are getting.

Here is a chicken coop made of recycled pallets.

Backyard Chickens http://ow.ly/huSH50CsG6e

Chicken coops made of recycled materials may not necessarily mean just using reclaimed wood. It might be re-purposing a free or cheap find on Craig's List. I love this chicken coop that was made from a play structure.

The Poultry Guide http://ow.ly/RHa850CsGb4

These people picked up an old shed (for free) and transformed it.

This is what they started with.

After some work and ingenuity...

This is the old shed transformed into a chicken mansion!

They tell you how they did it on Backyard Poultry http://ow.ly/onoZ50CsGp4

There are many creative fun examples of coops made with re-purposed items. Some of the ones that follow below use the re-purposed item as the run, and you build the coop, while others use the re-purpose item for the coop, and then you build the run. Let these ideas tickle your imagination.

A Kitchen Cabinet

A site called Inhabitat featured this http://ow.ly/YNiR50Csl4v

Has your kid grown out of their trampoline?

The Homestead Survival http://ow.ly/kwUw50CsIdm

How about recycled pallets and a water tank?

RecyclArt has a lot of fun ideas http://ow.ly/6uBJ50CsIiL

Kids grow up, and they do grow out of playhouses. Whether it be yours or a local find, these make some of the best coops.

The Thrifty Couple http://ow.ly/v40t50CuAOb

The link for the photo above has a lot of examples of repurposed playhouses. Here is a search of images of many playhouses re-purposed as coops. Feast your eyes and get some ideas; there's a lot of playful creativity here.[19]

The thing to remember if you are re-purposing is that you need to follow basic chicken coop protocol. Remember

- You want the roost to have a perch and be the highest point.
- The nesting boxes need to be dark and have built-in access to collect eggs and clean out the bedding.
- Make sure that the re-purposed item allows for enough space for the chickens you want **or** that you limit the chickens you have to the super cool re-purposed item you found. Either way, don't crowd your chickens.

If you search around for DIY coops, beware of misleading titles like "Build a chicken coop for $50 in less than an hour." Videos or blogs with titles like that will often teach you how to build a small run, not a coop with a roost and nesting boxes. Also, note when you have the instructions for a coop without an enclosure, you will need to purchase or build one unless you are in a rural situation where your chickens can roam free-range safely.

Chicken Coop Accessories

Emergency Back-up Coop

Whether you buy or build your coop, it is wise to have a small back-up coop ready if needed. A variety of situations can come up suddenly, and if that is so, the last thing you need is to have to figure out where to put a chicken or two.

You might use a back-up coop because:
- A hen might get sick, and you don't want to infect the rest of your flock.
- You might have a bully and need to isolate them away from the flock.
- If you have a rooster, you may want to give the hens a break.
- One of your hens might have wounds and need to heal. Wounds can be caused by a variety of things, such as being on the low end of the (literal) pecking order, suffering a predator attack, mating, or getting caught in a fence.
- You might have a hen hatching chicks who would like some time away from the flock.
- If you are hatching chicks, you may want to give them a little space on their own till they grow bigger.
- You might be introducing a new hen or two to your flock and want to give them a bit of space while they (and the flock) adjust gradually and integrate.

One of the DIY or re-purposed coops could be easy and inexpensive to set up and give you the option to use if necessary.

Feeders

Most chicken experts advise using hanging feeders instead of feeders flat on the ground. We agree. Incredibly, even just a little rise is not so attractive to mice (or worse, rats).

One piece of important advice: If you have six chickens, don't just have a feeder for six chickens. Ideally, have two feeders for six chickens each. Have food spaced out, preferably in several places, so that a less dominant chicken can easily access the food without being shooed away.

This footnote will provide you with a variety of feeders as examples[20]. One even has a "rodent-proof" feature.

Waterers

Chickens need a lot of water. Plenty of accessible fresh water is critical to their health and egg production. Many waterers can be attached to a water source so that you don't have to be filling it all the time.

Important advice about waterers: You have to make sure that you have several sources of water. As with the feeders, the less dominant hens need access without restrictions, and in the case of any automatic system, there may be a mechanical problem causing malfunction. Both access and back-up are essential for your chicken's access to water.

Are you in a cold climate? Even if actual freezing is not an issue, does it get cold enough to make the water unpleasant to drink? If so, there are stand-alone heaters you can buy as well as waterers with integral heating systems. Have a look at this footnote for options. [21]

Pecking Blocks with Treats

These treat-blocks, also known as "flock blocks," are convenient, but they come with pros and cons and some words of caution.

Pros: Chickens are entertained by pecking and scratching, and keeping them happy in this way decreases the chances that they will "act out" in bad behavior from boredom.

Cons: Like any animal, including us humans, chickens might forgo their feed for the treat-block. Treat-blocks are higher in fat, and they are not calculated to be nutritionally optimal for health and egg-laying.

Some chicken raisers put their treat-blocks out for limited amounts of time, like an hour or so in the afternoons. Still, others warn that they had experienced their chickens not eating their usual feed, waiting for the treat-block, and then going into a frenzy when it appeared.

One good use for a treat-block can be in the winter. In places that have frigid winters, the chickens may want to stay inside their coop (don't we all?). Putting out a treat-block for a few hours a couple of days at a time is a way to get them out of the coop and give them a little motivation to move and break up the boredom.

If you use a treat-block at all, the key is to mix it up so that its appearance is both limited and unpredictable.

Treat-blocks are expensive to buy; there are a lot of recipes out there. Go to the footnote for some recommendations.[22]

I love this way of giving healthy treats. This photo is from a forum on the backyardchickens.com website, which is an excellent resource for chicken experts. As a beginner, I always relied on them when I didn't have time to contact my local chicken experts personally. They are legitimate and will not spam you when you join.

Backyard Chickens forum http://ow.ly/RAR350CsIuC

Toys

Happy chickens are not bored. Yes, they have enough smarts to get bored, and that may lead to behavior problems.
Here are some options, *check the footnotes after the descriptions for videos and photos of playful chickens.*

- Agility Courses

Agility activities break up the boredom and also give you a way to interact with them so that they are tamer.

Here is a chicken agility course that this family put together; the young daughter leads the chicken along with treats. The hop through the hoop at the end could break the internet with its charm.[23]

Another video shows a set up indoors where the chicken is doing the agility course on her own without being led by treats. This example could be another winter "hack." You could set something up in a barn or covered space to give them some interaction and a break from the coop.[24]

- Swings

Chickens love to swing.

This page on My Pet Chicken has a swing and a video of the chickens using it (chicks, too!). Notice how they learn how to lean forward and back to swing themselves just like we do![25] I like this swing because it has a way for the chicken to hold onto with her feet. Beware of flat swings. I've never used a flat one, but I am highly skeptical of adverts out there for swings that show a chicken that is photoshopped on the swing.

Here is a DIY swing featured on Backyard Chickens with instructions and a video of a chick on the swing.[26]
Putting a swing in a brooder can be entertaining for both chickens and humans, it is highy recommended.

- <u>Fun scratching places</u>

You can offer up varied scratching places to relieve boredom. Keep them busy and engaged with piles of leaves in the autumn or piles of straw. One day, try replacing hand-fed treats with several piles of dirt in the enclosure that includes some of their feed, grit, and fly grubs. (Always offer two or more piles so that those lower on the pecking order get a chance to share in the fun.)

Dust Bath

Put this "accessory" in the category of "must-have" rather than as a "like to have." It is more like a feeder or waterer. Chickens need to take dust baths, and they love them. This is their way of cleaning and nature's method of ridding themselves of parasites. It is a primal need for them, so if you don't give them one, they will be stressed and seek to make their own.

An attractive dust bath is large enough to hold a few chickens at once (they like their spa time with their girlfriends) and at least an inch or two deep. Ideally, they love a bowl shape filled with dust to fluff their feathers. They will also roll. (This can be disconcerting the first time you see it if you don't expect it.) The dust needs to be fine and soft.

Generally, I like to have dust baths outside of the coop and inside the enclosure. Whether mammal or bird, any living being with lungs is vulnerable to respiratory issues if breathing dust in an enclosed area for an extended period.

Inside an enclosure/run, outdoor dust baths might be several small toddler sandpits filled with construction sand, fine dirt, and a little diatomaceous earth if you choose to add it. (Diatomaceous earth is discussed below when we talk about bedding.)

Outside an enclosure, you might want to make a larger dust bath made of dirt depressed into soft straw bales or a child's play-pool filled with dirt.

Inside the coop, something simple like a boot tray works great.

Amazon http://ow.ly/zFDd50CsIJD

Some people just put a pile of construction sand in the corner of the coop or run, make a depression and fill it with some fine dirt and forego the container. You can try that; it depends on whether you want to tolerate the mess.

Once you've decided where you want to put your dust bath (some people offer two to their flock – inside the coop and out), you want to fill your container with attractive dust bathing material:

- Building sand (Remember, do not use play sand made for sandboxes and child's play. Play sand often has salt in it, which is not suitable for your chickens.)
- Fine Dirt
- Wood Ash (if you have access to it). Make sure that your wood ash has been cooled for at least 48 hours. It's incredible how it can hide embers, and you don't want your chickens getting a fire lit under them!
- Diatomaceous earth (if you choose to use it)
- Optional: Add dried lavender or dried lemon balm – an excellent addition for the chickens and you too![27]

Mix it up and watch your chickens enjoy themselves. [28]

Egg Crates

You will need to get your eggs from the nest box to the kitchen intact. A simple basket might be enough, but if you have more eggs than a basket will protect, this is a great idea. These come in small, medium, or large, or you could easily make your own.

Lehmens http://ow.ly/nmRM50CsIMZ

Bedding

You've got your coop set up. Now consider what will make your chickens comfy and encourage them to snuggle down and lay those eggs.

I'll start by suggesting that cedar shavings are strongly discouraged. Yes, there are reports of people having no troubles, and cedar is attractive because it can kill parasites. The problem is that cedar oil contains elements that are toxic to chickens, especially their respiratory systems.
Speak to a knowledgeable chicken vet about this issue to get their advice if you want to risk this.

First, let's discuss bedding for the coop, then we will turn our attention to the flooring for the run/enclosure.

What are the functions of bedding in your coop?

- In the roost
    - Catching the poop. (Chickens poop all the time, including a lot at night).
    - Insulation.
- In the nest boxes
    - Bedding is there to be comfy, warm, soft, and the best feeling for the chicken to want to lay eggs.
- In the covered area of the coop
    - Bedding catches and absorbs the poop and makes it an excellent place to scratch and hang out if your chickens need to be inside for protection or don't want to be outside.

Typically, you will see articles and advice about the bedding to use for your coop. I separate the coop parts by function because you may choose a base bedding for the whole coop, but mix in extra absorption for the roost and "living room" and add in a little extra softness for the nest boxes. You don't have to, but it is essential to understand the *functions* of the bedding to be able to understand why one material is better than another.

Generally speaking, the main criteria for bedding is that it is
- Non-toxic
- absorbent
- quick drying
- compostable
- and relatively inexpensive.[29]

After those criteria, you need to decide what material will be the best for the happiness and comfort of your chickens. The absorbency and compostable nature of the bedding will make it more comfortable when you clean the coop.

Pine or Other Wood Shavings

You can get wood shavings from a local woodworker or purchase it from a feed store. It is absorbent, and the chickens scratch around in it happily.

If the bedding stays dry, it will not smell, but if it gets wet (from an overdose of runny poop or spillage from the waterer), then it will certainly smell. Generally, the absorbency of wood shavings means that they don't smell. It is one of the strong points of this bedding.

You may find out that you have a choice of "cut" size for the wood shavings. You want a larger cut for wood shavings as the smaller ones can be too fine and lean towards sawdust. Filling your coop with sawdust means that you can risk wood dust, which can cause respiratory problems in chickens.

If you decide to do the "deep litter" method of composting your chicken manure, wood shavings are the best choice as they have enough absorbency.

Some chicken raisers report problems with crop impaction resulting from the chickens ingesting wood particles. Others have used wood for years and never had this problem. Paying attention to the cut is the first line of defense against this. Consult a knowledgeable vet for their opinion.

Tip: If you can source a local woodworker or other business for your wood shavings, then you are more likely to get fresh-cut wood. This is much preferable (especially if you are composting with deep litter) because the wood still has sugars that are active in it. This process helps break down the elements in the poop and composts it quickly.

Hay or Straw

Hay/Straw: What's the difference? Hay still has nutrition/seeds such as alfalfa that is given to cows. Straw is the chaff that has no nutrition. Either are options for bedding as both have the same pros and cons. The only exception to that is that hay is not as cheap as straw.

Straw is usually easy to find, and it is free or cheap. This attribute makes it an excellent choice for a daily or weekly coop cleaning routine. It's soft, and chickens love to scratch in it. It is also warm and gives a lot of insulation, which can be a big plus if you live in a cold climate.

The main downside of straw is that it is not very absorbent. Straw is not a good choice for a deep litter method of manure compost and coop care; it needs to be replaced often. If it is not replaced frequently, the coop will end up smelling, and that ammonia odor is not just unpleasant but harmful to your chicken's health. Straw gets heavy and clumpy with poop and is not as easy to clean in any case.

If you choose straw, make sure that you source it from someone who can guarantee that there are no pesticides in it. The chemicals in pesticides could kill or sicken your chickens.

Seasonal Bedding Materials

Pine needles are not absorbent; they are not good for an entire coop's worth of bedding. They can, however, be mixed in, and it's a great way to compost them. I like to rake up pine needles after the snow melts and then pile it up to dry. Afterward, I add them to my coop bedding through the summer.

Dried leaves (they must be dry) are also a great source of soft bedding. Put them in a place where they will stay dry, and you can access them through the year.

You can include dried leaves and pine needles container (carefully) in a deep litter coop, but not used as the primary material. You would have to experiment and carefully monitor your coop to make sure that the compost was ok. Any smell of ammonia, and you know that you need some wood shavings to bring it back to balance.

Sand

Some people report success with sand – if you do your own research on this, please note whether the sand is being used in the **run** or inside the **coop**. That makes a big difference. I've not used it personally, but I've known people and seen reports on forums of people who tried it and had respiratory illness and freezing in their flock as a result. When it gets soaked, it is heavy and clumps with ammonia smelling chunks that are a hazard to your chickens. Moisture means that in the winter, it can also freeze and not just to be poor insulation, but an actual heat absorber.

Also, if you use sand, make sure it is **not** "play sand" or sandbox sand. Both are too fine and can cause dust and respiratory problems for the use of bedding inside the coop. They also sometimes contain salt, which is not healthy for your chickens except in measured doses.

A bowl or pile of sand in the run is an excellent choice for grit and scratching around.

There are some ready-made chicken coop bedding options available:

Koop-Clean by Lucerne Farms. It is a blend of chopped hay and straw with a mineral-based odor neutralizing ingredient called "Sweet PDZ." I've not used it, but I'm interested in mixing it with my usual coop bedding and might include it in the winter when my chickens and I are all less active. [30]

You can purchase **chopped straw.**[31]

Aspen shaving nesting liners [32]

Premium pine shavings[33]

Please stay away from the following two materials for your coop bedding:

1. Beware of Cedar Shavings!

While it is debatable whether it is ok to use cedar as wood for a coop, there is a strong consensus amongst chicken raisers that cedar shavings for bedding are not safe. There is a chemical in natural cedar oil that is toxic to chickens and can give them respiratory problems. Cedar is tempting because you may hear that it does kill parasites. That's awesome, but the downside is that it may kill or is likely to sicken your chickens too.

2. Sawdust

Sawdust is excellent for compost but not for bedding. It cakes and quickly absorbs too much moisture that doesn't dry out, leaving a stinky, matted mess that is difficult to clean and will soon smell of ammonia and be a hazard to your chickens.

In the Outdoor Run/Enclosure

If you are not using a chicken tractor to keep your chickens moving, you'll want some kind of material on the floor/ground of your chicken run. Otherwise, it will become a mud pit. Chickens and mud are not a good mix.

Wood shavings, straw or hay, can all work in a coop, but for the run, you can also add wood chips. Wood chips work fine for a chicken enclosure.

Mulch also works fine; you can lay down a layer of mulch, then on top of that, you can add your compost and garden waste, especially dried leaves.

A Bedding Accessory – Diatomaceous Earth (DE)

Many chicken raisers consider diatomaceous earth (often abbreviated to DE) to be a kind of secret ninja move for their chicken raising.

Diatomaceous earth is found at the bottom of ancient seabeds. It is the skeletons of the cell walls and shells of single-cell marine phytoplankton (diatoms) that have fossilized and, over time, have turned into powder. The DE is grounded further into a very fine powder. There is commercial/industry DE that is used for many uses, including home cleaning as a cleanser. The DE used in a chicken coop is a food-grade DE. This is an important distinction. DE is used as an insecticide for fleas and other insects, a de-worming treatment, and source of minerals.

The next page shows you a computer rendering of one of these lovely microscopic beings.

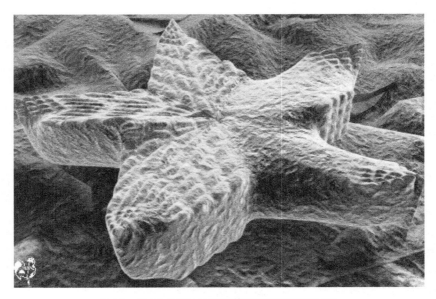

Raising Happy Chickens http://ow.ly/Uwao50CsJ7q

Chicken raisers use DE to exterminate mites and other poultry loving insects as well as helping with odor elimination. Chickens can ingest it, and it is claimed to be an effective worm treatment. (I've not found any studies to back that up, but chicken raisers say this is true.) Many chicken raisers rub it on the walls of the coop, sprinkle it in the bedding throughout the coop, and mix it into their dust bath and feed.

As DE is a very fine powder, you should wear a dust mask when applying it inside your coop as it tends to fly around, and you wouldn't want to breathe a cloud of it. (The same would be true of any fine dust.) Also, wear gloves as it can dry out your skin.

Concerns about DE

There are chicken vets and raisers who voice serious concerns about DE.[34] During the course of my studies, as far as I can tell, while some people have had problems, the data does not back up a need for banning DE is non-toxic, and its deadly impact applies to insects. If you get the right DE and apply the right amount, it will not harm you or your chickens. It contains minerals, and the food-grade DE is used as a mineral supplement for both chickens and humans. The manner in which it acts as an insecticide is not by chemical poisoning. It's mechanical.

If you put it under a microscope, you will see that those cell walls and shells from the ancient diatoms have tiny jagged edges. Being so fine, when it gets on an insect, it sticks and coats them so they cannot receive or maintain moisture. They dehydrate. They also get cut up. The jagged edges of the silica wear on the insect as it moves and literally shreds it.

For this reason, you don't end up with generations of insects that form a resistance to poisons. DE keeps working. It also does not lose its effectiveness over time.

Is all food-grade DE the same?
>**No**, there are not all the same. You want to ensure that it has no more than 2% crystalline silica (more can cause lung damage to you and the chickens) and its ground, 5-2mm.

Remember good protocols:
- Don't overcrowd your coop, or even keep it at maximum capacity.
- Ensure you have adequate ventilation

- If you have a dust bath in the enclosure inside the coop, forego the DE in that dust bath. Add it to the one outside.
- Apply DE sparingly to the walls but thoroughly in the crevices. Use a paintbrush. Sprinkle, and don't pour, when you apply it to the bedding and dust bath.
- Keep your chickens out of the coop when you are applying the DE and let them back in when the literal dust has settled.

This article explains both the concerns and the data for the claimed uses, which have been confirmed. She also points out the claims that are only anecdotal.[35]

I incorporated DE in my bedding, walls, and dust baths when I started raising backyard chickens over a decade ago and have always started with it "out the gate" when I've moved and set up a new coop. I've *never* had a mite or worm problem with my chickens nor lung issues.

Please note, my experience or that of others is not data; it is anecdotal evidence. I encourage you to read the articles in the footnotes about DE, including the scientific articles, to decide for yourself whether DE is the choice you want to make.

Tips:
- It's better to apply DE more often in small amounts rather than dump a bunch at once. Apply a thin coat on the walls of the coop with a paintbrush, making sure that it gets into the crevices of the coop, and "sprinkle," and don't "pour" in the bedding, dust bath, and feed.

- It isn't effective when it gets wet – keep your DE dry.

- Do *not* use DE inside your coop if you are working a deep litter system. The DE will kill the good bacteria that are essential for compost, and you will have an ammonia-smell-ridden coop that is both unpleasant and dangerous to the health of your chickens.[36]

Coop Bedding Conclusion

Remember what the criteria are for the bedding for your chickens and your convenience.
- The chickens need a place where they will be clean, dry, insulated, have plenty of space and can scratch and peck without getting sick.

- You need to find bedding and a coop cleaning protocol that is reasonable for your lifestyle.

The run has different needs as it is outside and does not require a comfy nesting space. I like a solid layer of mulch on the bottom. Then I maintain wood chip on the top.

You don't have to choose between the options discussed above; you can mix and match. My personal choice is pine shavings with straw mixed in for the coop bedding. The pine shavings are the best for absorption, but the straw adds softness and insulation.

The deep litter method can work well if you have the right ventilation, a spacious coop, and you do not have to use DE. It is not recommended for beginners unless you are very experienced with composting and know how to watch for the signs of healthy compost and are confident that you can maintain this. See the chicken manure compost appendix for more information.

Choosing Your Chicken Breeds

The first question you need to ask is, what do you want your chickens to do? Are they a source of eggs? Do you want them for meat? Do you want them as pets? Do they need to be child friendly? Do you want varied colors of eggs? Do you want both eggs and meat? Do you want a quirky showcase of unusual looking chickens?

There are three main categories, which are "laying hens" and "fryers," and those that are considered "dual."

- **Laying Hens**
  These hens give the highest yield of eggs.

- **Fryers**
  The chickens raised for meat are larger with more muscle

- **Dual**
  A lot of backyard chicken raisers opt for "dual-purpose" hens who lay at least a medium amount of eggs and are also on the large size. The advantage of this is that when a hen has completed her 2-year cycle of laying, you will then have a meat hen.
  - Note that you may find the word "dual" used on the internet to refer to both laying hens, who are tamed, gentle, and friendly. In this guide, the term "dual" refers to hens that are both laying hens and large enough for meat, as described above.

In all of these categories, almost every breed has another subset:

- **Bantam chickens**
  Bantams are the miniature version of any standard breed. Being smaller, they lay smaller eggs and are not usually considered meat chickens; however, being smaller means that they take up less space. Bantams are very popular in urban settings, and individuals and couples or small families find that they get plenty of eggs for their needs.

Bantams are not usually known to have any personality differences from their standard counterparts. One thing to note is that they are less suited to cold climates. Some are rated explicitly as "not cold tolerant." As we discuss breeds, remember that you have a choice between Bantam or standard size chickens.

Bantams fly, and they are good at it. Most outdoor enclosures require cover from aerial predators such as hawks, falcons, and owls. For Bantams, you have the added reason for a cover because they can fly away.

Make sure to provide Bantams with some high perches to fly to as well as their enclosed high roost. This is for two reasons: 1) they like them and 2) so they can get away from the standard size hens if by chance you are mixing them. Bantams can be vulnerable to bullying and pecking from the larger chickens.

Another question that is important to ask is, "Am I going to have any roosters?" Most urban and suburban municipalities will not allow roosters, so just keep reading about the breeds of hens. You need to have 10 hens per rooster. If you will not have at least 10 hens, then a rooster is not the right choice.

If you **are** planning to have a rooster, then we suggest starting with the hens if you are a beginner. Get used to keeping the hens, then introduce him later. If you are going to get a rooster now, then skip to the rooster section. In many breeds, the temperament descriptions of the hens are different from the roosters. If roosters are in your future, think carefully about the breed of the rooster(s) you want to manage and also whether you want full breeds or hybrids in your flock. If you only wish for purebreds, then the breed of rooster you choose will obviously dictate the chickens you get. If you want a different breed of chicken, do your research on the rooster of that breed.

Let's take a look at a curated selection of individual breeds of hens. Each chicken will have the same list of attributes that are commonly requested.

- Laying hen, fryer, or dual
- Egg size (sm/med/lg) and how many eggs laid per week
- Temperament with humans (note if especially **child friendly**)
- Temperament in the flock
- Egg color(s)
- Heat/cold tolerant
- Availability and Cost
- Special characteristics

After we discuss each specific breed, you will find the lists containing just the names of the breeds (no descriptions), indicating:

- Laying hens
- Meat hens
- Dual hens (suitable for both eggs and meat)
- Gentle and social with humans
- Especially child friendly
- Good influence in the flock
- Lays blue, green, pink, purple, or dark brown eggs
- Heat/Cold tolerant

Australorp

ChickenMag http://ow.ly/z9AR50CsJoJ

Dual: Australorps were bred to be dual production chickens in Australia. They are Australia's national bird!
Size: Large
Eggs: 5-6 large eggs/week
Egg color(s): Brown
Temperament with humans: Tame, friendly, especially kid-friendly
Temperament in the flock: – calm and easy-going. Not usually aggressive
Heat/cold tolerant: Cold-hardy
Availability and Cost: Common
Special characteristics: Australorps are not particularly broody, and chicken raisers sing their praises about their egg production and temperaments, with humans, other animals, and other hens in the flock. They are an excellent choice for a beginner backyard chicken raiser, but on the large side. So, you have to be aware of this if you have a small urban space.

Rhode Island Red

Dual: Like the Australorps, the Rhode Island Red was bred to be a dual chicken. There are [37]two statues honoring the Rhode Island Red, where the breed was developed.

Size: Medium

Eggs: 5 large-extra-large eggs/week

Egg Color(s): Brown

Temperament with humans: Friendly, especially kid-friendly

Temperament in the flock: Can be aggressive and bossy/dominant in the flock

Heat/Cold tolerant: Very hardy in all climates

Availability: Common

Special Characteristics: One of the top egg layers in the dual-purpose breeds. Great for a beginner in terms of health and dual production. Their temperament can be a little "all over the place." You will see them described as "calm" and "aggressive" in the same sentence. My own experience with them was to think that if I could have a flock of only one breed for dual production, I'd consider them. When I had them, I noticed that they bully other breeds of chickens and not theirs. The size was not a factor; the other chickens were at least as large. In the end, they became meat chickens, and I chose different breeds for the eggs I wanted.

Plymouth Rock (aka Barred Rock)

Claborn Farms http://ow.ly/w7H250CsJNn

Dual: Another US east coast chicken that was bred for dual production.
Size: Large
Eggs: 4 large eggs/week including winter
Egg Color(s): Light brown
Temperament with humans: Friendly, tame
Temperament in the flock: Not aggressive, mind their own business.
Heat/Cold tolerant: Hardy in all climates, especially cold tolerant.
Availability: Common
Special Characteristics: Since they are so easy to handle, Plymouth Rock is a popular chicken and an excellent choice for beginners. They are not prone to broodiness and are generally very cooperative. There's a reason that decades of farm kitchen décor has featured this classic black and white barred breed.

Speckled Sussex

Dual: This breed was originally developed in Sussex, England, for meat. Over the years, it has been bred to be a dual-purpose bird.

Size: Large

Eggs: 4 large eggs/week including winter

Egg Color(s): Light brown

Temperament with humans: Super gentle, kid-friendly

Temperament in the flock: Even-tempered, not bullies

Heat/Cold tolerant: All climates (cold tolerant)

Availability: Common

Special Characteristics: It is easy to confuse a Plymouth Rock with a Speckled Sussex at first glance, but let's look closely at the difference. The Plymouth Rock is black and white barred. The Sussex has dark burgundy/mahogany flecks in the speckles. The coloring of the Sussex is that the feathers are tipped with black and white, and the base is the darker red/brown color. So even though the Sussex hens are also primarily black and white,

60

there is that dark reddish-brown in their feathers. Also, even though both have a single comb, the Sussex's is larger than that of the Plymouth Rock.

The only thing about these hens is that they tend to be broody. Otherwise, I personally loved them as a chicken that is friendly with adults, children, other animals, and the rest of the flock.

Sussex chickens are smart enough to be curious and need to have some entertainment, so they don't get bored. They will even wander if they need to find their own fun.

Wyandottes

Wyandotte Nation http://ow.ly/K4qs50CsJZR

Wyandottes are another breed that has a dark base with white speckles.
Dual (they are medium-sized but stocky, so there is a lot of meat on them)
Size: Large and heavy
Eggs: 4 medium eggs/week
Egg Color(s): Cream color
Temperament with humans: They are not aggressive but are described as "aloof" in multiple resources. They don't want to be picked up and cuddled and won't jump in your lap or follow you around. Even so, they don't get upset easily, and in fact, are a favorite for 4H competitions.
Temperament in the flock: They are not aggressive when they are with other chickens, but they always try to be dominant. They are not bullies just for the sake of it but will be fierce if provoked. They will not tolerate being on the low end of the pecking order.   If

you want to have Wyandottes in your flock, choose less dominant chickens to surround them, and they are fine.

Heat/Cold tolerant: All climates, cold tolerant.

Availability: Common

Special Characteristics: If you want to breed your chickens, consider Wyandottes. They have the reputation of being attentive mothers. As you might expect, this means they are prone to being broody.

One of the big plusses of the Wyandotte is that they are quieter than many other breeds. They are an excellent choice for egg and meat production in more densely populated areas. If you live in an urban environment with a tiny space, then the Bantam variety of this breed might be worth considering.

Leghorns

My Pet Chicken http://ow.ly/SaDf50CsK52

Layer
Size: Medium
Eggs: 4+ large eggs/week
Egg Color(s): White
Temperament with humans: Not affectionate nor friendly. Very independent. Leghorns can be nervous and wary. They are not a relational chicken, and they do not want to be held and cuddled. They are not good for children because they can get scared easily. It is recommended that children be supervised while with Leghorns.
Temperament in the flock: They are not particularly aggressive with others in the flock.
Heat/Cold tolerant: Heat tolerant, not cold tolerant. The primary reason for their cold intolerance is that they have single combs that are so large that they flop over. These combs are vulnerable to frostbite. Some say that this can be mitigated by applying Vaseline to their combs and wattles. [38] Other than their combs, they are a very hardy bird

for cold or hot climates. You have to ask yourself whether you would want to have the extra winter chore of Vaseline's application to your chickens. You can also find Leghorn chickens with rose combs instead of the single floppy combs.

Availability: Common

Special Characteristics: Second only to the Isa Brown for laying, the leghorn chicken is often used for industrial egg production. If you have a small space in an urban setting, the bantam variety of this breed can give you a few chickens that lay a lot of eggs.

Dominique (aka Pilgrim Fowl)

The Chick Hatchery http://ow.ly/Teea50CsK8V

Fun fact: The Dominique chicken is the oldest breed in the US. They were brought here no later than the 1750s. We don't know whether they were passengers with the pilgrims or brought from Haiti (which was a French colony known as Saint Dominique).

Dual
Size: Medium
Eggs: 4 medium eggs/week
Egg Color(s): Brown
Temperament with humans: They are calm and docile. Gentle with children, so they are a good family choice.
Temperament in the flock: Not aggressive or particularly dominant.
Heat/Cold tolerant: Very cold-hardy.
Availability: Previously endangered, now numbers are rising because of more people raising backyard chickens. Dominiques are becoming more common; you are likely to be able to find them.
Special Characteristics: Tend to get broody.

Buff Orpington

My Pet Chicken http://ow.ly/yb7O50CsKcC

Dual

Size: Large

Eggs: 3-4 large eggs/week

Egg Color(s): Brown

Temperament with humans: Affectionate, tame — enjoy being held and petted. Kid-friendly

Temperament in the flock: They are so calm and docile that they can be vulnerable to being bullied if mixed with more dominant hens. Since they are likely to be on the bottom of the pecking order, don't mix them with breeds that tend to be dominant or aggressive like the Rhode Island Reds or Welsummers.

Heat/Cold tolerant: Very cold-hardy

Availability: Common

Special Characteristics: Great mothers but can be broody over unfertilized eggs. The mix of their large size and laid-back nature makes them susceptible to overeating and obesity. Make sure they get enough exercise, and avoid giving them too much food.

Jersey Giant

Fryer
Size: Largest chicken!
Eggs: 2-4 large eggs per week
Egg Color(s): Light/Medium brown
Temperament with humans: Very gentle and friendly. The only hesitance for children is that they can be so big and heavy, and that can make it hard for a young child to handle or carry them.
Temperament in the flock: They get on with other chickens, are not aggressive, and don't end up being on the bottom of the pecking order because of their size.
Heat/Cold tolerant: Not very heat tolerant, but very cold-hardy.
Availability: Common
Special Characteristics: They will go broody, including in the winter. They can be depended on to be patient and hatch their eggs if you choose to breed them.

Isa Browns

Layer – Bred for industrial egg production
Size: Medium
Eggs: 4-5 medium eggs per week, including through winter
Egg Color(s): Brown
Temperament with humans: Isa Browns are very friendly and affectionate with humans, and they are kid friendly.
Temperament in the flock: Many chicken raisers report that they are fine with their own breed but will be aggressive bullies with chickens of other breeds. They lay a lot of eggs and are gentle with humans, so if you just want to have one breed and want a lot of eggs, then the Isa Brown is a good one.
Heat/Cold tolerant: Definitely winter hardy, they are good tolerant of heat as long as ample shade and water are provided.
Availability: Common
Special Characteristics: Isa Browns were deliberately bred in the 1970s for factory farm egg production. Amazingly, they are friendly to humans, but one characteristic that has bred out of them is broodiness. This can be a plus if you want eggs and do not intend to breed your hens and hatch your own chicks. If you do want to hatch chicks, then consider an incubator and brooder as Isa Browns have no interest in motherhood.

Welsummers

Chicken raisers often choose Welsummers for their gorgeous looks and beautiful eggs. They are not a breed that we recommend for beginners. We've included them here so that you don't come across them and their eggs and don't have the details you need to decide whether you want to take them over.

Dual
Size: Medium
Eggs: 4 med/lg eggs per week
Egg Color(s): Dark terracotta brown with dark speckles
Temperament with humans: They are not aggressive, but standoffish and certainly not tame. They don't like to be handled, and they're definitely not good for children.
Temperament in the flock: If you are mixing breeds in your flock, they can either tend to be at the bottom of the pecking order or go to become bullies. Whether they are at risk for bullying will depend on which breeds you are mixing them with. When Welsummers are the bullies, they are not extremely mean, but they will lean to the bullying side with the less dominant breeds.
Heat/Cold tolerant: Cold-hardy
Availability: Rare but can be obtained via online hatcheries and are not very expensive.
Special Characteristics: They are on the quiet side, which makes them a good choice for suburbs.

70

Salmon Faverolles

Fryer: Bred in France to be meat hens
Size: Large
Eggs: 4 medium size eggs/week
Egg Color(s): Light Brown
Temperament with humans: A good choice for children; they are cuddly and like to relate to humans. Even though they are energetic and playful, they are very gentle.
Temperament in the flock: There is information out there that suggests that Salmon Faverolles should not be in mixed-breed flocks because they will end up at the bottom of the pecking order. If you dive into chicken forums, you will see others saying that they have had no problem. The Faverolles tend to be docile, perhaps giving way to the more dominant chickens, but not picked on.
Heat/Cold tolerant: Cold-hardy

Availability: Not as common as most of the others, but are becoming more popular. If you can't find them locally, you can find them on online hatcheries.

Special Characteristics: Faverolles are energetic and can be cartoonish in the way they run around, "like a chicken with their head cut off", but with their heads securely in place. Some describe them as very "vocal." Others say that they are not any noisier than the other breeds in their flock. This may depend on the individual, but no one describes them as exceptionally "quiet." They will go broody if you want to have chicks but do not have a strong tendency to brood constantly.

Easter Eggers

Steemit Homestead Section http://ow.ly/dxsf50CsKE8

Easter eggers are not a breed, and they do not have a standard. They are also not accepted in competitions. They are a hybrid of breeds of chickens that result in specific egg colors.

Layer
Size: Small
Eggs: 4-5 med/lg eggs per week
Egg Color(s): The color(s) of the eggs will depend on the hybrid you have. You may have blue, aqua, light green, olive green, pink, purple, lavender. Note: each hen lays one of the listed colors of egg. You won't have one hen who lays rainbows. If you want a mix of colors, then you will have to find a hen that lays each particular color.

Temperament with humans: Friendly to humans and are also kid friendly. Not just friendly, they like attention and handling and will even jump into your lap. They can be super fun as family backyard chickens because of the egg colors and their personalities.

Temperament in the flock: Not bullies, because they are small and gentle. They can end up being on the bottom of the pecking order if mixed with breeds that are more dominant or aggressive. If, by chance, you would want to mix them with others, then consider the gentler breeds like the Faverolles or Cochins.

Heat/Cold tolerant: Both heat and cold tolerant

Availability: Common

Special Characteristics: The amount of noise they make seems to depend on the chicken and the type of breed crossed in your Easter Egger. You will see people reporting that they are the noisiest in their flock and others saying that they want more Easter Eggers because they are so quiet. Curious and pretty energetic, so they need toys and activities to keep them from being bored.

Cochin

Fryer
Size: Large
Eggs: 2 sm/med size eggs per week, including the winter
Egg Color(s): Brown
Temperament with humans: Very gentle, known as "gentle giants." Excellent for kids, but maybe too big and heavy for younger children to handle.
Temperament in the flock: Not bullies, and because of their size, they are not vulnerable to being bullied.
Heat/Cold tolerant: Cold-hardy, not particularly heat tolerant. You need to pay attention to the heat. Provide plenty of shade and water.
Availability:
Special Characteristics: Feathered legs and are not suitable for mud. Make sure you have a mulch or some way to keep a dry run if you live in a "mud season" area. Cochins are bred and often obtained because they are incredible mothers. They will even hatch the eggs and raise the chicks of other hens who are not inclined to be broody. You will need to manage their broodiness when they are not hatching and raising chicks.

Brahmas

Pinterest http://ow.ly/Hv6g50CsKK6

Dual
Size: Very large - *almost* as large as Jersey Giants
Eggs: 3-4 Med/lg eggs per week
Egg Color(s): Brown
Temperament with humans: Very calm and gentle. Kid-friendly, but like the Jersey Giants, standing at nearly 30″ tall, and they might be a lot to handle for a child.

Temperament in the flock: They tend to be good with other chickens, a calming influence.

Heat/Cold tolerant: Both heat and cold-hardy *however*, they have feathered legs, which makes them not as suitable for wet/muddy environments, and care must be taken in the winter concerning their feathered legs.

Availability: Usually easy to find

Special Characteristics: Hens will brood on a nest and will tend chicks. They can go broody with unfertilized eggs but that tendency can be managed easily as they are so docile and cooperative when handled.

Polish

Pinterest http://ow.ly/fU7A50CsKO9

That is a real photo, it is not photoshopped. You can tell it's a Polish chicken by its headdress or "crown" that covers its eyes.

The name "Polish" is very misleading. We are uncertain of the details of this breed's origins, but it is most likely that the breed came from Spain to the Netherlands and that the word "Polish" is derived from the Dutch "pol," which refers to a large head. In any case, they are not from Poland.

Neither "layer" nor "fryer" nor "dual." The Polish chicken is a "looker."

Size: Small

Eggs: 2-3 medium eggs/week – but this is an average. People get Polish chickens because of their friendliness, as well as their fashionista headdresses, and not for their eggs. Some will lay more eggs per week, while others, less. The breed varied on this, and it is best not to rely on their egg production.

Egg Color(s): White

Temperament with humans: Very friendly, they make great pets for kids. They are easy to handle and will tolerate kid cuddling. Many people find them to be a delight as part of their backyard flock because they are friendly and a mix of both regal and comical.

Temperament in the flock: Will end up on the bottom of the pecking order because they are on the smaller side, and the headdress attracts bullying. Don't mix them with dominant or aggressive chickens.

Heat/Cold tolerant: Not cold-hardy. Part of the problem in the cold is that their crown can get wet and freeze. Super wet environments are not the best, either.

Availability: Relatively easy to get

Special Characteristics: If you want to participate in shows, then Polish chickens can be a lot of fun. Kids in 4H clubs often use them due to their cooperative temperaments. One thing to be aware of is that their crown limits their sight. As prey animals, if startled, they can easily get scared. Let them know you are coming, and teach your children to approach them gently and with warning as well.

"Bantam" Breeds

As we discussed earlier, most standard breeds have an equivalent Bantam variety. The following "Bantams" are either hybrids or do not have a standard size equivalent.

Silkies

Feather Lovers Farms http://ow.ly/OVzl50CsL8j

Fryer Size: Small
Eggs: 3 med eggs/week
Egg Color(s): White/cream

Temperament with humans: Wonderful pets, and extremely kid-friendly. Affectionate and like to be handled.

Temperament in the flock: They are vulnerable to being on the low end of the pecking order because of their pom-pom heads, small stature, and gentle nature. If you mix them with other breeds, ensure that you don't put them with aggressive or dominant breeds. Including a couple of Polish chickens can as well be a way to "normalize" the fluffy headdress look.

Heat/Cold tolerant: Both cold and heat hardy. Even so, their feathers are different from other birds, and when they get wet, they get soaked to the skin and cannot just shake it off as other birds can. So, even though they are "cold" tolerant, they are *not* "wet" tolerant. They have no insulation when they get wet, so you must keep them dry in precipitation to avoid hypothermia. They are also more work than other breeds if it is muddy.

Availability: You can usually find them locally or regionally. They have become popular because of their looks and friendliness as pets.

Special Characteristics: One look will tell you that a Silkie chicken is fluffy, adorable, and comical. Also, their plumage is like silk or satin. They are incredibly soft. Their feathers are different from other birds, and they cannot fly. You can find them in black and other variations as well. They are specifically raised for meat in China. An important thing to understand is that the meat and skin of the Silkie chicken is black, sometimes dark blue. In one place, that is a delicacy, and in other areas, it may be off-putting. The meat tastes like other chicken, but just a bit stronger in flavor without being "gamey." Silkies will get broody. Broodiness is great when you want to breed them, but you will have to manage their instincts if you only want the eggs.

Mille Fleur d'uccle (aka "Mille Fleurs")

"Looker" and Show Chicken
Size: Small
Eggs: 3 very small eggs/week
Egg Color(s): Cream
Temperament with humans: They are gentle and have an easy-going temperament, making them a great choice for a beginner chicken raiser. They are not, however, particularly affectionate. They can be handled easily and are "kid-friendly" in the sense of kids being "safe," but they don't necessarily want to bond or interact a lot. If you want a "pet chicken," unfortunately, you may be disappointed.
Temperament in the flock: Vulnerable to being on the low end of the pecking order. Be careful of which breeds you mix them with.
Heat/Cold tolerant: Mille Fleurs are heat tolerant, and they do better in the cold than many other Bantams, but not suited for extended zero/sub-zero F freezes. Also, they have feathered legs, which will need cleaning and tending in wet or muddy conditions.
Availability: You will probably find a breeder locally or regionally.
Special Characteristics: Unlike other "show" or "looker" chickens, instead of comical, Mille Fleurs are elegant. They are gorgeous birds; even their name means "a thousand flowers" for a good reason. They can and will fly. They do well in enclosures with netting or fencing on the top to keep out predators and keep them in. Mille Fleurs make good mother hens, but you will have to manage their strong, brooding instinct over their eggs. All chickens need space, but Mille Fleurs are particularly active, and they *especially* need enough space to avoid behavioral problems.

Frizzles

Photo by The Chook Coop on Backyard Chickens Coops http://ow.ly/xjvq50CsLti

Be aware that not all Frizzle chickens look like this one. Some have crowns; others are different colors. You can "Frizzle" any number of breeds of chickens, so they are not consistent.

"Looker" Show Chicken

Size: Small – Medium, depending on the mix

Eggs: 2-3 med eggs/week

Egg Color(s): Cream

Temperament with humans: One chicken video describes Frizzles as "docile lawn ornaments." [39] They are gentle and kid-friendly. Frizzles are often described as "sweet." They enjoy more human attention than Mille Fleurs.

Temperament in the flock: Like other show chickens, they can be bullied. They are gentle to other chickens as well, so you can mix them with other breeds just like them, such as Cochins, Silkies, or Polish chickens.

Heat/Cold tolerant: Their feathers do not give good insulation for either heat or cold. A moderate climate is most suitable for them.

Availability: Usually available locally or regionally.

Special Characteristics: Other "looker" chickens have been described as comical. Frizzles go beyond comic to utterly ridiculous, in the best possible way. They have been bred to have feathers that curl upward and stick out instead of lying flat like other birds. This gives them the "frizzle" appearance. Chickens of several breeds can be bred to be "frizzled," including Polish, Cochin, Silkies, and Plymouth Rocks. The "frizzle" comes from a specific gene that causes the feathers to form this way. Breeding a Frizzle with a non-frizzled breed will produce about ½ frizzled chicks and ½ standard feathering. It is not common to breed Frizzles with other Frizzles (called "Frazzles) because their feathers are not sturdy and will break, making them vulnerable to baldness. Frazzles can also have health problems.

Frizzles can't fly, so their roosting bars will need to be low to the ground. It also means that you have to ensure that you perform due diligence to protect them from predators.

Online Hatcheries

We've mentioned online hatcheries several times and wanted to provide clarity regarding the pros and cons of purchasing from them.

First, research the history and reviews from numerous sources, including backyard chicken forums, to find credible recommendations for hatcheries. You would want to receive healthy chicks that will survive and not bring illness or parasites to your flock.

Understand that the business of hatcheries is the mass production of chicks. You are not getting backyard chickens; they are factory farmed. If you have any interest in show chickens, stick to a reputable breeder as hatcheries do not care about maintaining breed standard specifications.

The benefit of hatcheries is that some claim that chicks are easier to tame than those hatched naturally and cared for by a chicken mother. If you find this to be an issue, then an incubator can help you to do the same thing.

Most backyard chicken raisers use a hatchery when necessary to find chickens of a breed that they cannot get locally. If you use this option, make sure you do your research.

Roosters

Whenever our current rooster strolls by, I roll my eyes and think, "it's a good thing you're so pretty, buddy boy." Crowing at a distance can be atmospheric. If he happens to start his daily routine at 4 am next to your house, it may be a different experience for you.

Earlier, we discussed hen bullying. The stakes are high if the hens are acting out and bullying in your flock. The stakes are even higher if you have an aggressive rooster. They can do a lot of harm to hens and even children or yourself! I've known more than one rooster who was sent to the stew pot for being aggressive with the hens, dangerous to a toddler, or being a menace to handle.

Besides temperament, size is also a consideration for your roosters. A large rooster in a flock of small or Bantam hens will likely injure them. On the other hand, a Bantam rooster with Jersey Giants will not likely be useful for hatching chicks.

We are focusing on the most well-behaved roosters here who are as suitable for beginners as a rooster can be. As with hens (and other animals like dog breeds), each breed will have personality *tendencies,* but temperaments can vary greatly depending on the individual as well as their environment and how they have been raised and treated. It is important to get a rooster that has been handled from an early age.

This guide is also working on the assumption that you are choosing *one* rooster for now. Most chicken experts recommend one rooster for every ten hens. Breeders will have more roosters per capita, but we do not recommend that a beginner start with a rooster at all, let alone more than one. Having more than one rooster requires a very careful analysis of the breed of rooster and the breeds of hens you have, what the dynamic will be between them, etc. It also crosses the line from backyard chickens to farming in terms of knowledge, skill, time, and attention.

Roosters are not just for beauty and breeding – they can also be great protectors for the flock. Occasionally we let the flock out on a sunny day for an afternoon nibble on the grass. I love watching our rooster follow the hens along. He watches over them so carefully and patiently as they feed, never losing his sentinel attention.

In general, roosters are more aloof towards humans than their hens. They tend to be pretty serious about the job they have to do. They want to mate and protect the hens. The roosters listed here are tamed well so that you can pick them up, but unless noted otherwise, they would not be described as affectionate. As always, individuals vary, and there are exceptions to the tendencies.

We have curated seven rooster breeds for you here. They are all good for beginners; you can look at the size and other characteristics and decide which one might be best for your flock and location. If you are not raising purebreds, you may also consider what the mutt chicks will look like when bred with your various hens.

I have also noted whether the rooster has feathered legs in addition to heat/cold tolerance. A chicken can be cold-hardy in terms of temperature, but with feathered legs, they can get frostbite and do not tend to be "wet hardy" in mud and rain. If you want a rooster with feathered legs, consider how you will mitigate the effects of rain, mud, and snow. It often takes daily cleaning and drying of those legs to keep them healthy.

Brahma Roosters

Brahma roosters can stand 2 feet high. Many smaller predators will be put off by a Brahma watching over your flock.
Get a Brahma rooster to mix with hens who are at least a substantial medium size and large sizes.

They are gentle with their hens and docile with their humans. Brahmas are *so* laid back that they can tend to be bullied by more aggressive chickens, so consider the temperament mix, especially if you have more than one rooster.

Brahamas are heat/cold tolerant; *however,* note that most Brahma strains have feathered legs.

Orpington Roosters

Orpingtons (aka "Buff Orpingtons as this is the most common color) are both large and heavy. Mix them with medium-large and large hens.

I am particularly fond of them as roosters as they are not only easy to handle but can also be friendly and relational to humans. They are gentle in the flock as well. On the flip side, they are known to be absolutely fierce when it comes to protecting their hens.

If your hens are large enough to handle their mating, Buff Orpingtons are both heat and cold-hardy and do not have feathered legs.

Faverolle Roosters

I've never had a Faverolle rooster myself but have had friends who have said that once they had a Faverolle, they never wanted another breed of rooster again.

They are large and can breed with medium-large hens.

Faverolles are elegant; they will be friendly and tame to their keepers and treat their hens well.

Faverolles are another breed that is cold-hardy with feathered legs; you will have to heed to the wet/mud/snow feathered leg warning. They are not particularly tolerant of the heat.

Plymouth Rock Roosters

My first chickens (including the rooster) were Plymouth Rocks. I loved them all. This is another popular rooster choice, especially for beginners.

Plymouth Rocks are a medium-sized breed – these roosters can breed with medium and some large breeds.

This is another rooster that will tame well with handling and be good to the hens. They don't seek affection with humans; they are focused on the flock and want to do their work. There is something about them that comes across as particularly centered and calm, and they have this effect on the flock.

Plymouth Rocks are both heat and cold tolerant and do not have feathered legs.

Black Australorp Roosters

Australorps are another good choice for beginners with other Australorp hens or other large breeds.

These roosters are quietly attentive to their hens; my Australorp was a keen observer and had a lot of focus on his protection. I've heard this from others as well.

They tame well for handling, but like the hens, you need to gain their trust and rapport. Once you do, they will be fine.

Australorps can overheat quickly because of their black feathers, so they are not a good choice for hot sunny climates. They are cold-hardy and do not have feathered legs.

Welsummer Roosters

Welsummers are medium-sized, and they will mate well with medium and few smaller hens.

They are a bit friendlier with humans when compared to a lot of other roosters. As well as being good to handle, they will enjoy your company. Not generally as a lap or cuddle rooster, but they will gravitate towards you to peck and feed near you if they are outside of their enclosure. This is an example of a breed where the rooster might be more docile to humans than the hens are.

You can't deny that Welsummer roosters are beautiful. Note that they do crow loudly and often.

Welsummer roosters are not the best in the heat but will tolerate it if they can access plenty of shade and water. They are particularly good in cold climates and do not have feathered legs.

Cochin Roosters

Cochin roosters are very large and imposing. You have to mix them only with large hens.

They tame well and are friendly, but unlike the Cochin hens, don't expect the rooster to be a lap chicken. If you have a flock of Cochin hens with a rooster, you'll have a lot of chicken love!
Cochin roosters are not heat tolerant. They are cold-hardy but have feathered legs.

Note that the Bantam Cochin roosters do not have the same temperament as the standard Cochins. Generally, Bantam varieties are like their standard counterparts, but in the case of the Cochin Bantam rooster, this is not true. The Cochin Bantams tend to be **very** aggressive with humans, other roosters as well as the hens.

Part Three: Maintenance

Now that you have your coop, your backup coop, and your chickens, it's time to get into the routine and go into "maintenance mode." We will look at the everyday needs, everything from feed to dust baths, and finally, cleaning to ensure a healthy environment.

Chicken Feed and Nutrition

Basic Nutrition Needs

Water

Don't make the mistake of making plentiful feeders and dust baths with only one water station. Ensure that there are plenty of waterers for your chickens and that they are fresh and abundant. The more dominant chickens will push others away from water and food. Any sense of scarcity will trigger this response.

Water is critical to every living being, but "If a laying chicken goes without water for more than 12 hours, it can go out of production for weeks." [40]

Like humans, if the water is not fresh and attractive, the chickens will not drink it, leading to dehydration. Figure out how you are going to supply plenty of fresh water. How you deliver this will depend on your lifestyle and environment. Are you away from 6 a.m. to 8 p.m. in a very hot or very cold environment? Will you be at home and be able to observe and tend the waterers a few times a day? Think about how you would describe your lifestyle and climate, and then consider what kind of watering system you need to provide for your flock.

If you are in a cold climate where the water may either freeze or be unpleasant to drink because it is so cold, you need to consider how you will heat your water source. There are options out there to choose from, and this is another place where local knowledge and experience can be a real godsend. You can make inquiries from your vet, feed store, or a local chicken expert. Local experts and feed suppliers can give amazing advice based on where your electrical source is, as well as the kind of coop setup you have. Sometimes the local feed stores carry equipment that is particularly well suited to solutions for your local environment.

Another feature of your waterers will be height. If you have a flock of mixed sized breeds, make sure there are plentiful waterers for the small hens as well as the large ones. You would want all of them to be able to reach it easily.

Chicken Feed

It is vital to understand the basic nutritional needs of chickens at the key stages of their life so that you can compare feeds when looking for the best possible fit. If you have chickens at different life stages together, you want to make sure that their life stage needs are being met.

There's a bit of vocabulary to understand about chicken feed. Let's look at some standard terms:

- Mash
  - Mash is finely crushed grain, protein, and supplements. It's almost like dust or powder. Some chicken raisers use mash for a while when their chickens show signs of aggression. There are a lot of opinions about whether it is good to use mash as a steady diet for all ages or not. It is common to use mash for chicks.

- Pellets
  - This is the most common chicken feed. The mash is taken and made into hard pellets, satisfying that primal pecking need.

- Crumbles
  - Crumbles are often used for baby chicks. It is just the pellets broken up into smaller pieces.

You will have the choice of whether to use medicated feed or not for your chicks. If there is any question about them being vaccinated, many chicken raisers do this for protection. Typically, I hatch chicks from my chickens or acquire chicks from a breeder I trust, so I just use organic chicken starter feed. A couple of times, I have acquired chicks from unknown sources, so I used medicated feed. When hatching my chicks, I get them vaccinated and feed them high-quality organic feed.

Like humans, nutritional balance varies with age. We will discuss the appropriate percentages of nutrients for the lifespan of the chicken.

For details of the specific micronutrients that chickens need, I would recommend these two readable, science-based articles from the University of Georgia and Alabama & Auburn Universities. [41]

Chicken feed is carefully formulated and measured with instruments and equipment that the average citizen does not possess. DIY chicken feed is not recommended. Don't risk the health of your chickens by trying to make your own feed. Choose a reputable brand for your feed, and then DIY your treats and pecking toys for snacks.

Adults 18+ weeks

The protein content needs to be 15-18%. It is essential to ensure that your chickens are getting enough protein as it takes a lot of protein to lay a good egg! It also takes protein to maintain healthy muscles for meat chickens.

Once the hen has started laying, the calcium now needs to go up to 3-5%. Calcium is required for healthy hard shells. Don't start this level of calcium until the chicken has begun laying so that you don't get a chicken with kidney stones or other health issues.

Note that if you used medicated feed for your chicks or pullets, you might need to wait for a while before you start eating their eggs. Follow the instructions on the feed you are using.

Which feed grind should I give to my adult chickens? Pellets, mash, or crumble?

You can feed any that suit you, actually. It is not required that you move your chickens to pellets. Adult chickens like pellets over mash because the mash is so small and powdery that they can't eat it as easily. I have found out that the best is crumbles. Crumbles also work well for adults of the smaller breeds or Bantams who struggle with pellets.

One factor to consider for feed is the mess. Mash and crumbles are messier. Wherever you choose to store your feed, set yourself up so that you don't spill when you are filling the feeders. Chicken feed attracts rodents. You can imagine that mash and crumbles are more likely to make a mess than pellets.

Another factor to consider is your humidity. Mash and crumbles are more prone to mold, and mold kills chickens. I've never had chickens in high humidity, but if I did, I would most likely choose pellets.

Gravity based feeders work much better with pellets. The mash and crumbles end up clogging and getting moisture in them and being prone to mold. If you choose mash or crumbles, don't use a gravity feeder. Mash and crumbles will make more mess below the feeder as well; this will attract rodents into your chicken enclosure.

If you have trouble with rodents around your enclosure where your feeders are, you can bring your food in at night. This can do a lot to discourage intruders.

Feed Storage

You have to think through how you store your chicken feed. The importance of storage cannot be stressed enough. Your feed needs to be kept dry so that it does not mold. I use a small galvanized-steel garbage can with a lid that fits tight.[42] I am very careful to scoop the feed into the feeder and put the top back on immediately, so I've never had my feed go moldy using these. Any sign of mold or possibility that your feed got moldy (somebody didn't put the lid on tight when there was a fog, rain, or humid day) would result in that entire feed being thrown out, and the container thoroughly cleaned and dried completely. As I said before, moldy chicken feed will kill your chickens.

Besides mold and humidity, another consideration is rodents. Chicken feed attracts rodents, so making sure that it is well secured *and* that you don't spill it when you fill the feeders is critical.

Ever since I've been using the galvanized-steel cans I mentioned in the above footnote, I have not had a problem with either rodents or mold. Set up your protocols to prevent moisture and rodents, and all will be well.

Grit

Do you know the saying that something is "rarer than hen's teeth"? The reason for that old-fashioned phrase is that hens do not have *any* teeth. As a result, they have evolved to digest their food without chewing.

The food they eat goes to their crop and stays there for a while. It then moves on to the gizzard, where the food is broken down for the rest of the digestive system. What "chews" the food so it can be digestible is grit.

Grit in the diet is essential to the health of a chicken. We are going to discuss different kinds of grit and how you might incorporate them into your feeding.

There are a lot of conflicting opinions and ways of incorporating grit amongst chicken experts, and it can be confusing. The first confusing thing is that oyster shell as a supplement is called "oyster shell grit." It is and isn't grit. Just because you're giving your chicken oyster shell (which is provided to layers for calcium) doesn't mean that you don't have to provide other grit.

There are generally three types of grit:

1) Insoluble grit
   This is ground granite or flint. This is the material that the chicken needs to digest the food well. If there is not enough grit in their gizzard, they will not digest well and even develop a "sour crop" where the food goes rotten in the crop because the gizzard does not have enough grit to digest the food. (Sour crop is discussed in Troubleshooting.)

2) Soluble grit
   This is oyster shell grit. Oyster shell is used as the best source of calcium for chickens. It is a kind of grit and critical to add to the diet of your adult chickens. It is not sufficient on its own as the only source of grit that is required for digestion.

3) Mixed
   You can purchase grit mixes for convenience.

It is important to get the right grind size so that it's not too big for the chicks or too fine for the pullets or adults so that it passes right through and doesn't do its job in the gizzard.

How to Give Your Chickens Grit

There are a variety of ways to give your chickens grit:

- Mixed in their feed
  - If you decide to mix it in their feed, then you will have to look at the instructions on the grit as well as that on the feed you are using. Some feed comes with grit already in it. How much you mix in will totally depend on the feed.

- A separate grit feeder or "hopper."
  - Here is an example of a grit hopper that you can have available to the chickens on demand. I like this one because it is intended for the outdoors and it's rainproof. Grit must stay dry.

Amazon http://ow.ly/LIXq50CsLDE

If you want to use a grit hopper, you will need to make sure that you have several so that the chickens on the lower end of the pecking order are not kept away from getting their grit. Just like the food, you will need to supply plenty of grit so that there are no questions of scarcity.

- Scattered on the floor of the enclosure

Some chicken raisers just scatter the grit around on the floor of the enclosure and let the chickens scratch, peck, and find it. Experiment with this and make sure that whatever grit you put out is gone in about 20 minutes.

Of these three options, my preferred choice is the grit hopper stations. The challenge of putting it in their feed is gravity. Grit is heavy and small, so it will not be evenly distributed as intended when it sifts down to the bottom. I like scattering it on the floor, but I have to think about if they have enough or too much, and whether it's going to get wet, and if it does get wet, then what am I going to do about it.

Although chickens will overeat and make poor nutritional choices if given too many treats, healthy chickens will self-regulate very well. They crave grit for their digestion and will not eat too much unless there is some dietary imbalance going on. When I make it available and plentiful, they just get it and regulate themselves beautifully.

Grit for Chicks up to 8 Weeks

Some chicken raisers say that you don't need to give any grit to chicks that are on a starter feed. The feed will have some in it, and they will do fine. Others say that they start spreading grit on the floor of the brooder so that the chicks can begin to access it as needed.

Don't give your hatchlings any grit until after 4 days. If you do decide to give your baby chicks grit, make sure it is a grind that is fine enough for them.

Grit for Chicks 8-18 Weeks (aka pullets)

As your chicks move to their next stage, they will need to start some grit if you haven't started already.

Make sure that you have a grind for this size/age chicks so that their bodies can use it.

The grit for this age might say "pullet" or "grower" grit.

Grit for Adults 18+ weeks

The adults can access the grit from the feeder, so you will need to have it mixed in the flock feed or spread on the floor of the enclosure.

Can free-range chickens get enough grit from just pecking around outside?

The answer is: It depends. If you are in a place where the soil is clay, and you have thick grass, then the answer will be, "probably not". On the other hand, if you're in the Rocky Mountains or the Sierras with a lot of granite, then. It will be most likely. One thing I love about providing grit in hoppers is that I don't have to figure that out. The chickens can get it if they need it.

Storing your grit

It is vital to keep your grit dry. I get a couple of medium-sized bags (because a large bag is heavier to lift and get to my shed) and put them both in a galvanized-steel trash can with a lid.

If you choose to use grit hoppers, then you will have to start experimenting to find out how much your chickens are using. Don't fill it to the brim at first; note how far it goes down over the course of a week. Then you can know how full to fill it every week. In the case of a particularly wet week, I fill it every other day instead of once a week, even though the hoppers are rainproof. The moisture still has an effect.

Treats

We discussed the option of treat-blocks in Part One under the accessories you can have for your coop. We need a balanced diet for our bodies, and we need to be aware of a balanced diet for our chickens and not encourage obesity or nutritional imbalance with too many treats.

On the other hand, treats are the spice of life! They also help with taming your chickens. Here are some treat ideas that you can use to feed your chickens with a few treats by hand. Children often enjoy getting the option of throwing treats on the ground or feeding their chickens with a few treats by hand.

You can turn some raw food scraps into hand-fed treats, such as chopping up some apples or carrots. Here are some you can buy to keep around for taming and pampering your flock.

Meal Worms

Chickens love mealworms. You can get them at your feed store or chicken supply. They are a great source of protein, and you can feed them by hand or on the ground. Mealworms have been my "go-to" treat choice for taming my birds, along with some "chicken crack" that I'll tell you about in a moment.

Treat Mixes

Have a look at these – chickens love them, and they are great for their health. [43]

Fly Grubs

When I think about treats like mealworms or fly grubs, I have to remind myself of "chicken aesthetics." It sounds gross to me, but the reason that chickens keep down the flies in your area is that they love the larvae. Give these as a treat, and they will love you for it.

Amazon http://ow.ly/BGWp50CsLGl

Chicken Crack

This is my favorite and my chicken's favorite. (The name says it all, really.) I rotate this in with the other treats listed here, but this is the one I run for if I need to catch a chicken or distract a rooster.

Amazon http://ow.ly/4YJZ50CsLL1

Chicken Crack mix includes bugs and a couple of grains. It also comes in sizes appropriate for different life stages so that your pullets don't have to try to eat the bigger pieces that the adults prefer.

Mealworm and Sunflower Treat

You may choose to get mealworms separately from your seeds and make your own mixes, but this is a very convenient, healthy, and attractive snack for your chickens.

Amazon http://ow.ly/2nyT50CsLMX

Seeds

You can give your chickens common seeds and nuts, but these two are the top treats. Both are presumed to be raw and unsalted:

- Sunflower seeds are a favorite with chickens. You can feed them with the shell or just the seed.
- Pumpkin seeds contain an ingredient that is a de-wormer for tapeworms, and chickens love them.[44]

Feeding scraps

One of the best things about backyard chickens is that your food scraps become your food in the form of eggs and meat. Also, you are keeping material out of the landfill and lowering your carbon footprint.

Scraps are not as convenient as the dry chicken treats, but they are very useful. They are still regarded as treats. They are a supplement but do not replace your feed, and you need to be careful about the amount you are giving them. Chickens are wonderful at eating up things like watermelon rinds and banana peels. In Part One, we discussed the treat feeders you can purchase or make as accessories in your coop. Here is a reminder of the treat feeder you can make for your raw vegetable scraps.

Before you feed your chickens with your food scraps, it is advisable that you check with your local ordinances pertaining to backyard chickens. Some municipalities have made it illegal to feed chickens with human food scraps in an effort to lower the risk of contamination from chickens to home kitchens.

There are a few food scraps to avoid:

- Mold or any spoiled foods
  - Avoid moldy grains, especially
  - Any spoiled food can produce toxins for chickens.

- Scraps with elements that are toxic to chickens
  - Avocado pits and shells
  - Potatoes or potato peels
  - Chocolate
  - Dried beans (cooked are fine)

- A lot of greases
  - A little bit of oil that was used to saute' some vegetables in some leftover dish is fine, but chickens are not made to digest grease, so deep-fried items and other greasy foods can be harmful.

- Garlic, onions, or a lot of fish
  - The taste of pungent foods like these will come through to your eggs.

- Raw Meat
    - The meat won't hurt them to eat, but it can trigger the cannibalistic tendency that you don't want to encourage.

- Processed foods or soft drinks
    - Give your chickens whole home-cooked food and vegetable scraps, not processed foods. These things are not healthy for us, and they are worse for them.

- Salt
    - A bit of salt in your dinner cooking is fine, but don't give them pickles or super salty food as too much salt is not good for them.

Note that we mentioned potatoes on the "no" list – but *sweet potatoes* are wonderful for your chickens – including the peels. Sweet potatoes are not nightshades and do not contain the toxins of concern; they are just packed with great nutrition.

Cleaning the Coop

Cleaning the coop has come up in previous sections; here is a convenient summary.

Clean the poop out of your coop every day. If it looks like a little bedding needs to be added to the roost or nesting boxes, then add a little.

Every week, you have to make sure that you change the bedding and compost it with your chicken manure. (Described below)

I recommend that you do not start with the deep litter method. Get your routine down doing the regular cleanings before you add in another learning curve.

As discussed, the "deep litter method" is a way of composting your chicken manure in the coop. This minimizes cleaning the coop to every six months.
Pros
4-6" of dry wood shavings will last six months.
It is an excellent method for winter in cold climates where you have snow accumulation and acclimate weather.
You only have to clean the whole coop every six months!

Put the compost straight into your garden beds; it's ready to go directly from the coop.

Cons

You can't use DE with the deep litter method because it will kill the good bacteria that is composting the manure. You'll end up with an emergency clean.

It's much harder to clean the surfaces of the coop to eliminate mites.

If you don't get the balance of manure and wood shavings quite right, it will smell terribly of ammonia, and instead of being easier, it will be an emergency health issue to your chickens. Ammonia will harm their respiratory systems.

Chicken Manure and Compost

What should you do with all that chicken manure? The best thing to do is compost it for your garden. The easiest and most ideal solution is to make a compost pile outside that starts with a thick layer of dry leaves or pine needles and small sticks with some fresh cut greens from grass or bushes. Put the chicken manure on top of that and cover it with the dry leaves or pine needles. Continue layering the dried "browns" (which have a lot of carbon) and the "greens" and chicken manure (which provide nitrogen). Usually, the ratio is 2-parts browns to 1-part greens/manure. You may need to save up your leaves and dried trimmings in the autumn and keep them dry to get you through the wet season. You might as well be able to get some from your neighbors.

This pile will slowly turn into wonderful compost for your garden. If it's too much for you to use, give some away to a neighbor or neighborhood garden!

Collecting Eggs

Collecting eggs (and eating them!) is one of the top joys of raising backyard chickens. It is a wonderful activity with kids.

The top reason for doing some daily coop cleaning is to keep your eggs clean, as we've discussed above. Let's assume you are doing some daily cleaning.

There are a few tips to help you keep the nesting boxes clean:

- Make sure there are enough nest boxes with the right amount of space and that they have plenty of soft clean materials. Chickens who venture out to make their own nest boxes can end up laying eggs where they may get contaminated or broken. Generally, experienced chicken keepers recommend one nest box to every four hens. You will have to be a watchful observer of your flock. If it seems like more space would help your hens to be happy in their nesting boxes, then add more.

- Never let your chicks or chickens sleep in the nesting boxes. Chickens poop a lot, including a lot at night. From the very beginning, when you acquire a chicken or bring a late-stage pullet/young adult into the coop, make sure that everybody has to leave the nesting boxes for the night. A chicken sitting on her nesting box pooping all night and then laying eggs is terrible hygiene and a setup for contamination.

  A couple of design points make nesting boxes less attractive as nighttime roosts:

  1) Make sure your nesting boxes are lower than the roosts.

  2) It is advisable to build nesting boxes with roofs to discourage the chickens wanting to roost in them.

- Collect eggs at least twice a day so that they do not sit in any poop or be pooped on. If possible, collect them when you hear her announcement that she has laid an egg for you.

- If a hen goes broody with fertilized eggs, don't let her hatch her chicks in the nesting box. You will have to move her. We will discuss this whole process in Part 5: Breeding Chickens.

- Keep your chickens clean. Poop on feathers can get on the eggs. When you do your daily look-over, check for soiled feathers. Sometimes you will have to bathe a chicken.

- Consider adding a "nestbox rollaway." Nestbox rollaways make the eggs roll away from the nesting box and into a separate container so that the chicken won't poop on them. This is one of my favorite recommendations, especially for people who are away during the day. The one shown here is 17".

105

Click image to open expanded view

Cleaning your Eggs

When and how to clean your eggs is hotly debated. The only thing agreed on is that at some point before cooking, your eggs will need to be cleaned.

I've heard so many different opinions on this subject over the years. When I make my own decisions, I like to mix a blend of science and common sense. Fortunately, now some voices have common sense and explain the washing issue clearly as well as basing their opinions on real data.

Let's start with a bit of important egg anatomy. When an egg is laid, the shell is very porous. It is encased with a membrane called the "bloom." The bloom is there to protect the egg from bacteria entering.

Washing eggs removes the bloom. If you are going to store your eggs, it is recommended that you do not wash them until you are ready to cook with them. You don't want to damage the bloom and risk bacteria getting into your eggs.

Should I throw away eggs with poop on them or wash it off?

Even this question is not straightforward. Quite a few chicken experts say that if it has any poop at all, it should be thrown away. An egg has 7000 pores. The risk of sucking bacteria into the egg when you try to clean it seems pretty high to a lot of people. Knocking the poop off with a dry brush can also remove the bloom, exposing the egg to risk. This chicken expert points out that a lot of the conflicting information about what to do has to do with your personal risk tolerance. [45]

On the other side of the spectrum, this chicken expert tells you how to clean an egg that has poop on it. [46]

While whether to clean or throw away an egg with poop on it is debated, throwing away an egg with a crack or hairline fracture is not. **Always** throw away an egg that has even the smallest crack.

Should you wash it off? Not if you are going to store it. Personally, I agree that if there is a lot of poop, it should be thrown away. If it's just a few flecks, then I store the eggs (we discuss storing below) and clean with this egg cleanser when I am about to use the egg. [47] The cleaning action is enzyme based with no chemicals. The ingredients are water, yeast, citric acid, and potassium sorbate.

How do I wash my eggs before cooking them?

Towards the beginning of this guide, we mentioned that you should never wash your eggs in cold water. If the water is cooler than the temperature of the egg, then when the water removes the bloom, those 7000 pores will suck in any bacteria that is on the outside of the egg.

I agree with this chicken expert[48] on the importance of water temperature because we know scientifically that the egg sucks in bacteria when water is introduced on the outside that is colder than it is, and the bloom is removed. The water needs to be 90 degrees F.

As I said earlier, I recommend the egg cleanser.[49] Keep it at room temperature and spray the egg as directed.

If you use water instead of just a rough brush, don't wipe the egg off. Let it air dry.

If this sounds confusing, that's because there are a lot of different opinions, and as noted earlier, you have to make decisions based on your own risk tolerance.

At the end of this paragraph, there is a footnote to several credible resources with different angles on this issue. Do your research, make your decisions, and then set up your system. Once you make your decisions and set up what you want to do, it will not be complicated or time-consuming. [50]

Storing your Eggs

We have established that, however you wash your eggs, they should not be washed until it is time for cooking them.

Storing your eggs might seem like a "no brainer": "put them in the fridge," right? Maybe not.

Eggs that are purchased from the grocery store have been washed with chemicals like bleach. The bloom is not there to protect the egg, so commercial eggs need to be refrigerated.

If you follow the recommendations we've made and do not wash your eggs until you are ready to cook them, then your eggs will not be as fresh after two weeks but will be safe to eat for three weeks. If you wash your eggs before you store them, they must go in the refrigerator.

For many people in France, Britain, and most of Europe, eggs are kept at room temperatures. There have been scientific studies done to determine whether refrigeration is necessary for eggs. The studies showed that *"there is no difference in the bacteria levels of cold storage eggs versus room temperature eggs".*[51]

My guideline is that my eggs can be on the counter in temps not exceeding 70 degrees for more than two weeks. The reason I have backyard chickens is for amazing eggs; after two weeks, they are just "eggs." Some chicken experts with larger flocks claim that they have an extra storage place that is cooler, such as a basement that is not refrigerated but keeps the eggs safe for two weeks and even more.

There are a lot of options for storing eggs on a counter, have a look here[52] (one is even cast iron with detailed molding). If you want to search for one, try using the terms "egg sorter" or "egg skelter."

Trimming Claws, Spurs, and Beaks

We are going to discuss trimming claws, spurs, and beaks in that order.

If you have ever trimmed the nails of a dog (cat nails are much smaller), then you will probably know most of this information and be confident in trimming nails and spurs the first time.

How often will I need to trim claws, spurs, or beaks?

In a natural environment, chickens and roosters are roaming free and usually have enough hard rock and dirt to be scratching and pecking in, so they do not need trimmings. When they are enclosed or kept in a soft grassy yard, there may not be enough hard stuff to trim claws naturally. Also, some locations have more rock and hard dirt or sand than others. If the substrate is very soft, the nails will not trim as well as it would if they have a lot of rock.

The best answer to the question is, "when they need trimming." Be attentive to your chickens. When you check their feet every day, observe how long it takes for their nails to grow and need a trim.

Trimming Claws

The right length for a chicken's claws is when they are on the same level as the toe. When they start to grow longer, they begin to curl. If untended, the chicken will have trouble walking.

Chicken claws are like that of other animals. They comprise of the same material that our nails are made of, and like other animals, they have a "quick" in the middle that is a blood supply.

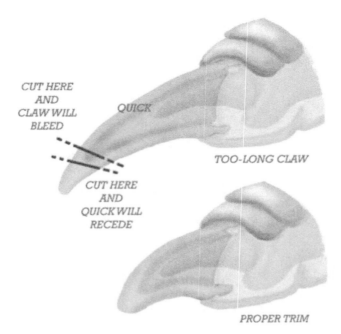

CUT HERE
AND
CLAW WILL
BLEED

QUICK

TOO-LONG CLAW

CUT HERE
AND
QUICK WILL
RECEDE

PROPER TRIM

The quick grows along with the nail. You don't want to cut the nail below the end of the quick because it will cause bleeding and pain. The illustration above clearly shows where you can expect the quick to be. You should be able to see it when the chicken's toes are clean.

The easiest way to trim nails is to gather your trimmer and file alongside a bucket with warm water. Get the chicken in a towel and put the chicken's feet in the warm water to soften the nails and wash the feet. After a couple of minutes, clean and dry the feet. You should be able to see the quick clearly now.

The key technique to use when cutting is to trim with several minimal cuts. Don't try to find the quick and cut at the edge of it in one cut. Cutting this way may make the nail split beyond the quick, cause bleeding, pain, and be a concern for possible infection while it heals. Keeping a chicken foot wound clean is a challenging prospect. Best to cut about 1/8" at a time to ensure that you don't split the nail.

You want the claws to be even with the end of the toe and pointing straight forward. The photo below of the rooster's spurs has a clear example of an overgrown toe. Scroll down and look at the right foot in that photo. The third toe has overgrown to the point of curling. You don't want this.

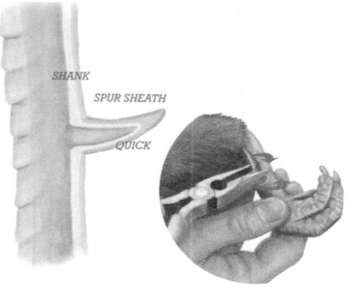

Backyard Poultry http://ow.ly/TedU50CsM0d.

When you are done cutting, file the rough edges of the claw. This will save both skin and clothing when handling the chicken.

After the nail is cut, the quick will recede. If your chicken's nails have become overgrown, you may need to cut a bit back, then do it again after a couple of weeks when the quick has receded. Repeat until the nail is at the right length, then maintain a healthy claw length.

Trimming Rooster Spurs

Roosters have spurs for fighting off other roosters, predators, and dangers generally. When the spurs get too long, they can be a menace to the rest of the flock in aggression and even to the hens when he is mating. Spurs can be very sharp, strong, and generally intense!

Rooster spurs have similar anatomy to the claws. You have to ensure that you do not cut the quick, and if the spur has grown way too long, you will need to do more trimmings at shorter intervals until the spur is at the right length.

Make sure that the tool you have is very sharp and that it's big enough for the spurs of the rooster you are cutting. For some roosters, large or very large dog nail trimmers will work just fine. For others, tools from your toolbox might be a better or necessary choice. Ask your vet or the employees at your local feed store if you aren't sure whether you have the right tool for your rooster.

Trimming Beaks

I want to clear up one thing about beaks immediately.
- We are talking about *trimming* so that the bird can peck and eat easily, *Not*
- *De-beaking* which is a cruel practice in battery farming to keep the birds from cannibalizing themselves, their eggs, and other chickens.

Like claws, the beaks of chickens would usually be trimmed by their natural pecking. This is not always the case in a domestic environment, and even chicks can begin to get an upper beak that is overgrowing the lower. It can get long enough to inhibit their feeding and natural preening, dexterity for moving objects, and social interactions. See the illustration below for an example of an adult chicken with her upper beak overgrown.

Backyard Poultry http://ow.ly/M79050CsM4i.

To trim the beak, gently but firmly hold the chicken's head, as shown in the photo with the chick below. With larger chickens, you may need a towel. With chicks, you can use a large nail clipper for humans, as shown below. This might work for your adults, but you may need a sharp large dog toenail clipper.

Just like the claws, you will need to clip just a tiny bit off with each cut. Only trim the upper beak so that it meets the lower beak.

Keep an eye out on your adult chickens. Many chicken raisers never have to trim the upper beaks of their chickens, but some do. When you catch the overgrowth early, it makes it easier for both you and the chickens.[53]

A Well-Ordered Pecking Order

"Pecking Order" is a chicken-raising word that has found its way into our general cultural vocabulary. Even a person raised in the city and living in a high rise who has never seen a chicken in real life knows exactly what the pecking order is. It is an order of hierarchy that is established by pecking. The corporate world has made a great deal of use of this highly applicable metaphor.

It is essential for chicken raisers to understand that the pecking order in itself is natural and healthy. Establishing a pecking order helps them feel safe and know where they stand. It is a primal response to flocking, so you cannot train your chickens to become egalitarians.

Our discussion here will focus on flocks that are comprised of hens only. If you have a rooster, then the rooster will be in charge, and the dynamic is very different.

When you first set up your flock in their coop and enclosure, they will go through the process of establishing the pecking order. If you have chosen your breeds thoughtfully in terms of personalities, this process should go smoothly.
The pecking order establishes who gets food, water, and dust baths first, as well as the choice and turn of nesting boxes and roosting space on the perch. It will also dictate access to toys.

The dominant hen is not just there to be dominant. The dominant hen has duties. She takes on the role of the protector in the flock that a rooster would have if he were around.

She will find a particular feeding place or go to the feeder first, but she will allow the others (in the order that they have worked out) to feed while she watches for hawks or ground predators. If the chickens are free-range, the dominant hen takes the role of finding the best place to scratch, find goodies and show everyone else.

When you first put your flock together, they will have to work this out. You'll see the hen who believes she is dominant fluffing up and sometimes squawking to let the others know she's boss; she goes first. If there is more than one hen who feels she should be dominant, there may be some fighting. (See the Troubleshooting section about bullying for managing this.) Keep an eye on these establishing dynamics. As long as no chicken is getting injured, it is acceptable and expected. Often, it's just a lot of noise and fluffing with some harmless pecks that don't draw blood.

The "middle" hens are working this out in their own way, but it isn't as obvious or dramatic. Still, keep an eye out for any pecking or injuries so that you can respond if any blood was drawn. (See Troubleshooting for wounds and bullying for the importance of dealing with blood right away.)

Once the flock has established the pecking order, things will go more smoothly if you've chosen breeds that tend to be less aggressive.

Introducing New Hens to Your Flock

Let's say you have your flock of hens who have figured out their pecking order and that you just acquired another couple of hens to add to the flock.

They are going to have to go through the process of establishing the pecking order all over again. There are some things you can do to make this a transition that is as smooth as possible without any hens getting hurt. This advice is not a magic bullet. Sometimes hens don't merge into a flock for reasons unknown to us. The important thing is to do everything you can to support the success and manage the new birds' transition or possible rejection.

Set up for Success

- Choose your breeds carefully. Think about who you are mixing with your flock and how you might expect them to fit into the current established pecking order. Are you adding smaller, more submissive hens to a group of larger hens? Are you adding a larger or more dominant breed?
  - Both size and looks are a part of the pecking order. If you get a hen that has a feature like a big headdress, she might get picked on by the other hens just because she looks different.
  - As a beginner, we strongly suggest doing your first introduction of new hens with birds about the same size and don't look very different.
  - It also helps if you choose birds who are around the same age. If you have a flock of hens who are toward the end of their laying years and you introduce young adults, that can cause drama. The same is true if your child fell in love with an older hen, and you put her in with younger laying hens.
  - Any factor that will make a chicken stand out, whether it be size, age, or looks, is a magnet for, at the least, not fitting in easily in the pecking order or, worst, for bullying.

- Introduce at least two at a time. Don't put just one new chicken in with the rest of the flock. It is best if you got two or more, and they had time in the extra coop before being integrated into the rest of the flock. At least they will know each other. It's similar to being a new kid in grade school.

When you have chickens to introduce, put them together in your extra emergency coop that is next door but not attached to the coop with the flock. Let them be in there together for a week; watch the reactions. If your current flock is one breed of chicken and you are introducing more breeds, then you might need to keep the newcomers two weeks next to each other.

Judge this yourself through careful observation. The first time you do it, you may not know, so just give it two weeks. During this time, the new birds will be able to sort their status between themselves, and the rest of the flock will get used to their presence without having access to peck at them. The chicken brains of the flock will wire the new hens as "not a threat" and normalize their presence.

The time of year varies with location and weather for the best times to introduce new chickens. I have never felt like the dead of winter with 2 feet of snow is a good time to introduce chickens, and I try to avoid "mud season" as the snow melts before there is any green growth. One time to avoid is universal: Do not introduce new chickens during molting season. There are enough stresses and challenges. See the molting section for more details.

Move your Chickens into the Flock[54]

When it's time to move the new chickens into the flock, put them in when it's time for your chickens to roost. Let the flock go in and settle on their roosts; when they have settled down, calmly bring in the new chickens, and they'll find their roosting space on the perch.

Set an alarm to be there the next morning when the chickens are getting up so you can supervise their daytime introductions to each other.

Make sure you have added feeders, waterers, or dust baths to ensure that there is ample access. More than enough. It helps greatly to put in a lot of new fun stimulation the first days of the new hens. Hang several cabbages on ropes for pecking, put in another raw vegetable treat station (or two!) Another recommended toy is a bale of hay or a pile of leaves. A sudden influx of new things will help the natural pecking order get established while also distracting the current flock from their focus on the new hens.

I had a mentor suggest to me that the next time I introduced new hens, I should move my coop. I tried it and will always do this if possible. In my experience, this made a huge difference. It breaks up the "my territory" tendency of the flock because everybody is getting used to a new space. Not everyone can do this, but if you have a chicken tractor, then this is a time to move it.

If you can't physically move your coop and enclosure, then try moving things around. Redecorate and move the furniture! Remodel the kitchen! Putting a hay bale, a pile of leaves, or new sand to scratch breaks up their space and changes it. You can put it where a feeder used to be and move that feeder somewhere else. Adding and changing the placement of toys will also give the coop a sense of newness so that the sense of change is not just focused on the new hens.

Another trick is to give them free-range time if you can. You could let them out in the morning onto the grass and let them have some time away from the enclosure where the current flock is likely to feel more territorial. Then bring them back into the enclosure with all the changes and distractions.

The First Week

It usually takes about a week for the new hens to be incorporated into the flock. There will be drama in the form of squawking, fluffing up, flapping, and pecking. There's no getting around this; it comes with the chicken territory.

When I introduced hens to a flock, I found out that handling and examining them twice a day instead of once was useful. Maybe it makes them feel more secure to have the uber-boss (the human) establishing dominance and protection. In any case, it gives me a chance to look for any blood from pecking. There have been a couple of times where there were no wounds in the morning, but by 2 pm, there was a little blood. If I hadn't checked in the afternoon, the wounds would have been left for the remainder of the day and overnight. By morning, that little bit of blood could have turned into a severe injury. Be especially observant at this time.

Caring for Chickens in Cold and Hot Conditions

We discussed some of the severe weather issues in earlier parts, but here it is in one section with more detail.

Chickens in Cold Conditions

Best Chicken Breeds for Cold Climates

If you are in a place with cold winters, having chickens that are naturally well suited to cold will save you a lot of work and possibly grief.

Generally, you want to choose larger breeds with small combs and wattles and lots of thick feathers. Larger combs and wattles are prone to frostbite. Some breeds can stay warm with their plumage in the cold, but it's just their wattles or combs that need tending. If this is the case with a chicken breed you choose, then you'll have to apply Vaseline or some other protectant to the combs and wattles to avoid frostbite.

Remember that feathered legs can get muddy if you have a mud season and will also clump and be wet in spring snow. While the feathered legs are a huge plus for keeping a chicken warm, they also require attention to ensure that they aren't a liability for the health of the chicken. Keep those legs clean and dry; they can get frostbite if left cold and wet.

Out of the 19 breeds we have covered in this guide, the ones that are best suited to be cold-hardy are:

- Brahmas (have feathered legs)
- Australorps
- Salmon Faverolles (have feathered legs)
- Buff Orpingtons
- Plymouth Rocks
- Wyandottes

Those are six awesome breeds to choose from; the only one that may not suitable for a beginner is the Wyandotte.

It is worth noting that Frizzle chickens are cold **intolerant**. They *look* fluffy because of the way their feathers curl out and up, but those feathers are for show and undermine any insulation for either cold or hot.

Coop setup and Care in the Winter

If you live in a place with cold winters, you most likely have a list of things you do in the autumn to prepare for the winter months ahead. Add your chicken coop prep to the list.

Check your coop for being wind, rain, and snow proof. Chickens are outdoor animals, and the cold-hardy types survive with shelter, but make sure that the wind, rain, or snow is blocked. They can lose a lot of heat through their legs; keeping the wind out makes a tremendous difference in keeping them warm enough. Even though the wind needs to be blocked, there needs to be ventilation so that you don't get condensation, which in the end, will make them colder and also encourage mold.

We do not recommend any heat lamp for a chicken coop, even one supposedly made for that function. One reason is safety (and an electrician confirmed this to me), while the other reason is that if you have a power outage, you will have chickens in your coop who have adapted to the temperature from the heat lamp. They are not equipped to adapt when the temperature suddenly plunges.

How will you keep the waterers from getting frozen or the water too cold to drink?
- You need to heat your waterers, but they must be checked regularly to ensure they are in working order. Also, check that they are working after any power outages. Check out the examples in this footnote for heating their water.[55]
- I live with very cold winters and therefore heat my waterers. I had an electrician come out and discuss the best way to do this safely. I keep the water inside the coop and check it in the morning and at night. I stop the automatic filler because we have to disconnect all hoses, so I have to make sure they have enough waterers so that I only need to fill them twice a day.

Dust Baths

Move one or two small dust baths inside the coop and make sure that others are under the covered area of the enclosure.

Cleaning the Coop in the Winter

Cold hard winters are a time when a deep-litter method makes sense. If it works well, you spend less time every day out in the elements, *and* the deep litter gives off heat to help them stay warm in the coop. I've known people who swear by it for the winter and early spring months, then give their coop a deep clean in the spring and go back to a daily/weekly cleaning for the remainder of the year. I've known two other chicken raisers who tried the deep litter method. They tried it in the winter and ended up having an emergency deep clean in two feet of snow and single-digit temperatures.

Deep litter is highly preferable during the winter months. We suggest that you practice it in the summer months. Go back and read the instructions, give it a try and get good at it. That will set you up for success in the winter months. Remember that you cannot use DM in your coop with a deep litter system as it will kill the good bacteria that is composting the material, and you'll end up with an ammonia smelling mess that is hazardous to your chickens.

Consider access to all your equipment so that it is easy, and all tasks in the elements and snow accumulation (if applicable) are the most effective.

The Outdoor Enclosure

You may have extreme days when the flock hunkers down in the coop, just like we do on a blizzardy day. They won't be in there all winter though; there's a reason we have the phrase "feeling cooped up."

If you set up your coop in the spring and put a tarp up as a sunshade during the summer, it will not be durable to provide protection during the winter. The chickens will need a covered area for their food and dust baths in the winter. You will have to find a way to get an A-frame or Lean-to roof that will shed the snow and give your chickens some outdoor space.

Make sure their dust baths, food, toys, and extra scratching are in the covered area. Boredom will lead to irritability and bad behavior, even more so when they can't spend as much time outside.

Another great thing for your chickens in the winter or wet/muddy season is to add some logs or other items for them to step on to keep their feet out of the cold and wet. [56] I learned this trick a few years ago when the article footnoted was published, and it was a revelation! The chickens *loved* them, I would brush off the snow and rearrange them, and it was part of their boredom breaking. It looked to me like they were using it as a kind of game or agility course. It was breaking the winter boredom for all of us.

Nutrition in the Winter

Your chickens will be working hard to stay warm, so it is appropriate to give them a little more feed. It is good to talk to your vet and local chicken experts about your area and your upcoming winter for recommendations. Some "cold winter conditions" are longer and colder than others, so how much more will vary for your individual situation. High energy treats and supplements can be an excellent idea for the coldest months, and if you can add them as treat toys to ease their boredom, all the better.

Broody hens in the winter

If you have a rooster and end up having a hen who is brooding on fertilized eggs, then move her to your brooding/hatching area. Her care will be the same but even more critical in the cold winter months.

For those hens who brood on unfertilized eggs in the winter, catch it early. This chicken expert advises that you treat a hen as broody if she stays in her nest box for more than an hour.[57] In the Troubleshooting section, we discuss details of broodiness and how to care for broody hens and break them out of the cycle if they are not hatching. Special considerations for winter are:

- Since she is eating and drinking less, she will lose weight and tend to get dehydrated. Dehydration can be a matter of life and death at cold temperatures since she cannot regulate her body's heat.
- She needs to be "broken" out of the broodiness cycle, not left to be.
- Some of the remedies, such as frozen peas, cannot be used as they may be life-threatening.
- The best thing to do is to put her in a wire cage with wire on the bottom. Read this excellent article specifically about caring for broody hens in the winter.[58] Catch the trouble quickly and bring her out of her broodiness before she is endangered, and the rest of the flock "catches" it.

Planning will be your best friend for a comfortable winter with your chickens. With a little forethought and preparation, they will fold into your winter routine and provide some eggs and fun entertainment.

Chickens in Hot Conditions

In general, smaller chickens with larger combs and wattles are best suited for hot weather. Some of the best, like the Silkies and Frizzles, have feathers that allow more airflow. The standard chickens that don't have unusual feathers will molt and grow their feathers according to the weather.

Out of the 19 breeds we covered in this guide, the chickens best suited for hot climates are:

Plymouth Rock
Speckled Sussex
Orpingtons
Welsummers
White Leghorn

A lot of Bantams do well in the heat.
Frizzle Cochins
Silkies
Easter Eggers
Millie Fleur d'Uccle Bantam

Caring for your Chickens in Hot Conditions

Always pay attention to their behavior. Just like us, it is natural for them to slow down, but they can also get overly lethargic and miserable.

Water

Water is the first consideration. You may need to have more waterers or fill them more often. Keep a close eye on this.

Coop Setup

Ventilation is critical for them to be able to take the heat. Coops should be built in a way that will enable them to block out drafts but also have adequate ventilation, so they are fit for both winter and summer conditions.

In climates where it gets hot but cold is not an issue even in winter, some chicken raisers put small windows in their coops to be able to get a cross breeze. If you do this, you can also set up a small battery-operated fan to move air at night if your chickens are miserable.

Once in a heatwave, I set up a pop-up shade structure over the coop; it made a big difference in the temps in the roost and nesting boxes.

Another trick is to use Velcro to attach some silver reflective material using the foldable shades for car windshields. This, too, can have a significant effect on the heat inside.

If you have an enclosure where 1/3 or ½ of it is shaded with a tarp or other covering, you may want to extend the shade space and put all of the dust baths in the shade if that's not too crowded.

Note how the sun moves in the sky during the hottest weeks, as you may need to change the angle of the shade cover to accommodate the summer season's sun angles.

Staying Cool

Besides water access, here are some fun tricks for helping your chickens stay cool:

Flick little sprinkles of water on them with your fingers – they perk up a bit and come running to me when I do this.

If you have a lawn in a dry climate, you can turn on a gentle sprinkler. You don't want to make mud, but the chickens love to be able to have some time to go in and out of it at their own will. I don't recommend this except in arid climates where they will dry out. In humidity, they won't be able to dry completely, and that can cause either getting cold at night if temps fluctuate or encourage parasites.

If your chickens are very lethargic, you can get electrolytes to put in their water.

Freeze treats, like watermelon rinds, for them to have cold stuff to peck on – they LOVE THIS! Just make sure that you bring out several – enough so that they don't have to fight for access. It must give them some relief because they can get on the border of frenzied excitement.

Conclusion of Caring for Chickens in Hot and Cold Conditions

The key to happy chickens and happy humans is for you to think through your needs in advance and plan for it. In the case of winter, there are more setups before the season. In the case of summer, if you have figured out how to provide shade to your coop beforehand, then it is handy, and you won't be running errands to get materials and building or assembling your shade structure in 103 degrees.

Besides planning, be a keen observer of your chickens. When the weather is changing, watch for signs of discomfort.

With planning and observation, both winter and summer can go smoothly and become a part of your routine.

Molting

I was doing so well with my first backyard chickens. I purchased them as chicks in the spring, had a learning curve, and learned to find chicken forums on the internet. I got through both summer and winter, and by the first year, I was in the groove and felt like I had this wired. They had gone through molts as chicks and juveniles; I thought that the adult molts would be similar.

Then their adult molt began.

I had vaguely registered that adult chickens have their first molt when they are about eighteen months old. In my imagination, this looked like more feathers on the ground, and perhaps some feathers would be sticking out at odd angles.

I was not prepared for the alarming sight of my chickens losing so many feathers, nor was I prepared for the sharp drop in egg production and my usually tame and social Plymouth Rocks being grumpy, distant, and looking miserable.

After a phone chat with the vet, a visit to the feed store, and talking to my chicken mentors, I realized that I had missed this piece of chicken education. I needed to take action to support my flock.

Based on their age, chicks molt three times before adulthood:
1-6 weeks
7-9 weeks
12-13 weeks

I was warned that my chicks would molt three times, and perhaps because they were small and growing, it wasn't frightening. (There is more information on this in the section on breeding and raising chicks.)

Adult chickens start molting around 18 months in the autumn. Their molting is triggered by shorter days, so the exact weeks they start molting will depend on your latitude. After their first molt, they will continue to molt once a year.

There is a particular cycle with an order for molting. Feather loss and replacement will take 8-12 weeks. *Sometimes* as long as 16 weeks, but that is not typical.

What is happening in the molt is that the new feathers have been forming under the skin of the chicken. When they are ready to come out, they push the current feathers out and emerge as "pin feathers."

Homestead and Chill http://ow.ly/55zB50CsMcw

It is important to know that unlike fully formed feathers; pin feathers still have a blood supply for their growth and establishment. If they are cut, they will bleed. The order of molting starts at the head and works down the neck, breast, body, wings, and lastly, tail.

Chickens will not only produce fewer or no eggs during their molt; their entire metabolism slows down. They eat less and poop less. Don't be concerned when you see their combs and wattles fading and becoming smaller; this is normal. Just ensure that they are getting enough water as well as the right nutrition, and all will be well.

How to Care for Your Chickens During A Molt[59]

Nutrition and Feed

When you notice that your chickens are starting to lose their feathers, you will also notice a drop in egg production. Some stop laying eggs altogether. This is a sign to begin to shift their nutrition. They need less calcium and more protein. Egg production takes protein, but feathers are comprised of 85% protein, so growing and replacing all of their feathers requires a protein boost.

- You want a feed that is 20% protein with great vitamins and minerals. Many raisers use a starter feed for this purpose. Ask your vet or feed store what they have in stock and that which they recommend. The critical component is that the protein content should be 20%. Don't make the switch suddenly – know the signs of early molting and make the transition over seven days.

When it's time for them to settle back into their egg-laying, don't change the feed suddenly. Transition carefully from the high protein molting feed to your usual feed for their best egg-laying health and happiness. Take about 7-10 days to gradually make the switch, replacing the molt feed with the layer feed a little more each day.

As always, your chickens need fresh, clean water, but when their bodies are re-growing all of their feathers, it's even more critical to their health.

High protein treats, *especially* grubs are great for their nutrition and their temperaments at this time. I like to use the Fly Grubs that we discussed in the treat section. [60]

It is important not to overdo it on fatty treats like sunflower seeds; a few are ok, but during this time, their bodies are craving protein, and they are less active, so too much fat content is not what they need.

Reduce Stress

- Molting is stressful for chickens, don't add any challenges.

- Do not introduce new chickens to the flock at this time. Having to figure out and re-establish the pecking order when they are grumpy and uncomfortable is a recipe for disaster.
- Molting is not a comfortable experience for your chickens. It is both itchy and painful. Even "lap chickens" will want to keep their distance. Reduce the amount of handling to the minimum for health checks.
  - It is important to handle them enough for mite checks. Their skin is uncomfortable enough. Watch their behavior for scratching, and hopefully, you did a deep clean of the coop in anticipation of the molt. Parasites can be more miserable during the molt.
  - Their skin is hypersensitive at this time, where the new feathers are pushing the old feathers out. Teach children that this is a time when it hurts them to cuddle. They need space to rest and grow out those healthy new feathers.

- Clean bedding will make them more comfortable if you usually change it over once a week, give your hens a bit of pampering by tending the bedding every day.

- Don't do a home improvement project on your coop or move it unless unavoidable due to damage or unexpected early winter preparation. Making big changes will increase their stress, do your best just to leave them with plenty of water, food, dust baths, toys (they may or may not use), and healthy treats.

- Chickens can get very grumpy at this time, so watch the dynamics. Since the pin feathers will bleed if cut, any pecking can cause even more blood than normal. As always, isolate any hen with a wound until she heals, and there is no more blood. Continue to provide healthy, pecking treats and toys to alleviate boredom. They may not use the toys, or one may use one to let off some grump steam instead of taking it out on her hen friend.

Part Four: Chicken Dinner

Your hens will lay eggs for two years, and then you have to decide what to do with them.

- Some raisers keep them as pets. The challenge is that as your chickens' age, you will have fewer layers and a whole lot of older pets.
- Some raisers send them off to local families or farms where they will be butcher for food.
- Many chicken raisers deliberately raise some meat and "dual-purpose" chickens. When the hens finish their laying years, they are killed for dinner or the freezer.

Killing a Chicken

If you've never killed an animal for food before, it can be very intimidating to think about it. I was raised in Los Angeles and never saw a chicken outside of a plastic package until I was a teenager. My parents were raised on farms and learned to do this from their parents, aunts, and uncles. Fortunately, I had local mentors who walked me through the slaughter process, in person, at my side.

These days, we may or may not be able to access a mentor who can be with us in person to go through this. One of the main reasons that people want to raise backyard chickens is to get away from the battery and factory farming practices that are cruel and inhumane, so they care about killing their them with as little distress as possible.

There are many ways to kill a chicken and a lot of conflicting information out there on the internet and in books. If you ask around, you may have conflicting opinions from your vet and local chicken experts. You are strongly encouraged to read this advice as well as talk to your vet, feed-store employees, and local chicken experts about what they recommend. Gather the information and decide what sounds best to you.

We will discuss one that is humane and, in my opinion, the most suitable for beginners. This is now my chosen method because I find it to be the easiest on myself and the hen.

Killing a Chicken with a Pellet Pistol

A lot of chicken experts say that using a gun is less humane than a hatchet because of the risk of missing. That is true if you are not point-blank. It would take a sharpshooter to dispatch a chicken reliably. Even from a short distance, you would be likely just to injure and terrify them and the rest of the flock.

What has worked best for me is to take the chicken *away from the flock* so she can't be seen. Wrap her up in a towel and kneel down with one knee, stabilizing her body to the ground and one hand holding her neck to the ground.

Place the pistol at the base of the skull (examine where this is before the moment you do it). Aim it diagonally upwards towards the area between the eyes, *not* straight across the face or at the jaw.

When you shoot it, it will both sever the spine from the brainstem and go through the brain. The chicken will die instantly and not feel pain. There will be a kind of jerking/convulsion of the wings, which is just the nervous system reaction; the chicken is not conscious; she is dead.

Here is a link to different kinds of pistols.[61] You can order one online, but I suggest going into a gun shop, tell the store owner what you are doing, and try different ones kneeling down in the position you are going to use it. I have relatively small hands, so my priority was to get a gun that I could hold and competently use for this purpose. If you are used to handling guns or have big hands, then the only question would be which gun would be the easiest to get the angle right on the chicken's head.

It's never easy the first time; we've all been there. Talk to other people and decide on the method that makes you feel the most confident. I will say that after using a hatchet a few times, and then a cone and knife twice, I settled with a pistol and haven't looked back.

Using a pellet gun is very reliable and fail-safe if you practice a couple of times on a towel or dummy. You can also practice picking up the chicken and holding her down in the position a couple of times before you do it. In any case, it is wise to have another pellet handy if needed, but it is doubtful that this will be required, just best practices.

Preparing your Chicken

Phew. Now you don't have to worry about suffering and trauma; just taking care of the meat swiftly.

Hose off the chicken to get any blood, dirt, or poop off of the outside. This will make the de-feathering process a better experience.

Now you need to get the feathers off. I was pleasantly surprised to find out how easy this is. Scalding will release the feathers.

- o Have a large pot big enough to fit the whole chicken boiling with water that is 150 degrees. Use a thermometer to ensure that the water is at this temperature.
- o Put in a couple of squirts of non-toxic dish soap. This helps loosen the feathers. If you're not using something with ingredients like fragrances, bleach, or other chemicals, it won't hurt to get it on the chicken at this point; it's going to get rinsed off. I like to use Seventh Generation Free and Clear for this purpose; it makes the de-feathering easier.
- o Submerge the entire bird using a pair of tongs or another implement to lower it and hang onto it. Stir it around. The temperature of the water is low enough so that the bird won't get cooked (keep checking the temp periodically to make sure it isn't going up) but hot enough to release the feathers. After about 50 seconds, pull on a feather and see if it comes out without pulling the bird up out of the water. If it does, then it's ready. If not, it needs a few more seconds.
- o Put the bird on a cutting board and pull the feathers out. There are such things as "chicken pluckers" that will do the job in 15 seconds. This one costs $460.00.[62] I can assure you that many generations have been processing chickens with just a big pot and a sharp knife, but if this looks like a worthy time and mess saver, then it is an option for you. It is a wonderful tool for farms that process multiple meat chickens for their family all at once, just like this one.[63]

The rest of the process is the detail of cutting off the feet and removing all the guts. It's more complicated than cutting up the whole chicken that comes with a packet of neck, liver, and gizzards in it. This site[64] explains it well with perfect photos so that you can see what he is doing. If you have killed and plucked the chicken as described above (our recommendation), then start at Step 07. This is precisely how my mentor taught me to do it. [65]

Part Five: Breeding Chickens

If you can have a rooster, you may decide to breed your hens. If you can't have a rooster and want to do some breeding, you can get fertilized eggs for your broody hens to hatch and raise.

Breeding Behavior

If you have a rooster, mating usually happens by letting nature take its course. The rooster starts fluffing up and prancing around the hen. The hen lies low for the rooster to mount her, and the mating is done quickly.

Sometimes roosters will have favorite hens. Those hens may get scratches and pecks on their backs and heads that lead to wounds. Isolate her and let her heal. Chicken mating is rough, but If this is a troublesome rooster in that regard, get some advice from a chicken expert or vet about your particular rooster and hens, and then you may decide on whether he goes to the stew pot. You'll get another rooster who is a bit more gentle on your hens.

If a rooster is not breeding, or if all of the hens are rejecting him, he might be sick. Take him to the vet.

Hen Care During Egg Development

Whether you have fertilized eggs from your own rooster, or you got them from a hatchery, now you need to give some special care to your hen who is sitting on her nest.

Moving the broody from her nesting box can be a little tricky but much easier with the right preparation. Set up her private brooding space before you move her so she and her eggs can be placed in and left without fussing with the food, water, and environment. Do it at night when she's most settled. I highly recommend using a towel. Her primal instinct is that being moved is wrong and dangerous. Be calm, gentle, but firm. If possible, do this with another person; you get the chicken, your friend gets the eggs, then you put the hen on her eggs when they are all in place.

Eggs take 21 days to hatch. It is best to have a broody nesting box away from the flock. It needs to be dark, clean, have a comfortable temperature, and protected from predators.

During this time, she needs to have all her self-care next to her; otherwise, she will neglect herself to the detriment of her health.

- She needs to have food and water very close to her
  - Since she is not producing eggs now, she will be eating less and does not need the same nutrition. My mentor suggested feeding the broody chicken with starter feed. I've had success with that and have found out that other chicken experts encourage this as well.[66]
- Space for her to poop away from her nest but still close to it
- Provide a small dust bath just for her.
- It is essential to keep the nest and private coop clean. Bacteria can kill the embryos.
  - Observe your broody and note when she leaves her nest to take her dust bath, eat, and drink. It is usually around the same time every day. This is a great time to get in there and clean the coop well.

Many chicken experts advise having an incubator on hand in case the hen starts pecking her eggs or goes off her brood. If you have invested in fertilized hatchery eggs, this may be an especially good idea so that you can be quick to intervene.

Congratulations! Hatched Chicks! Chicks up to 8 Weeks

There's nothing like baby chicks. Starting at day 19, start listening for peeps and watching closely for signs of hatching or any signs of chicks who have died not making it out of their shells. Also, watch the hen closely. There are occasions when a hen will reject her chicks. You can rescue them if you have a brooder on hand to take over from her.

If you have given the hen starter feed during her brood, she will be used to it, and the chicks will just start eating the same thing.

All chickens need a lot of protein; the baby chicks require 18-20% protein to build their bodies and get a strong start in life.

Calcium is essential later in a chicken's life, but too much in a baby chick's life can cause trouble such as kidney stones. No more than 1% calcium in chicks until they are laying. Do not give them an extra oyster shell.

Look for "starter feeds" that come in "crumbles" instead of pellets as the baby chick's beaks can't deal with pellets yet.

The hen will naturally teach the chicks to eat, drink, and bathe. You don't need to provide grit.

After about a month, it is advisable to start giving the hen some layer feed in order to support egg production. She naturally starts to distance herself from her chicks after 5-6 weeks.

Chicks 8-18 weeks (aka "pullets") and a Transitioning Mama

The protein content goes down at this stage. Feed them with 16-18% protein content.

There is a specific "grower feed" for this age. You can purchase one of these feeds or help the pullets make the feed transition. If you want to do it yourself with the feed you have, begin to move from the starter feed only to a mix of textures: keep some crumbles, but start adding in mash and pellets.

Introducing Pullets or Young Adults to the Flock

If you are purchasing chicks or pullets, it is essential to consider the breed. If you have a mixed flock, then you will just be thinking about personalities, but if you have a flock of one breed, don't use the young ones as the avenue to diversify your flock. The older ones may reject them entirely.

With pullets, introduce at least three instead of two. The little ones need each other to get through this transition, and numbers are powerful in the chicken world.

Put the chicks or pullets in the emergency coop or a wire cage next to the flock's enclosure and coop. Let them be there for *two weeks* instead of just one like the full adults. (See the section on Introducing Adults.) If they are very young chicks (not pullets yet), then you can put them in a wire cage like a dog kennel inside the enclosure or next to it. [67]

Put them into the coop as the chickens are just about to settle down to roost. Then they will wake up, and the chicks are "still there" instead of being suddenly introduced.

Like the adults, letting them all out to free-range in the morning, if you can, is a way to get the flock away from their enclosure where they will feel more territorial.

If you find that any of the hens are particularly hard on the little ones (older hens can tend this way), then put those hens into the extra enclosure for a couple of hours a day to give the young ones a break and let the rest of the flock adjust to them without the bullies.

One chicken expert suggests giving the little ones "an escape route." Put cardboard boxes in the coop with small cutouts where the chicks can get out of harm's way, but an adult is too big to fit.[68] This is brilliant.

As the young ones come into their adult bodies, they will start laying and get more confidant. The flock will figure out how they are integrated, and they will blend in.

Using an Incubator and a Brooder

Personally, I like the broody hen doing the work of caring for the eggs and raising the chicks. Using an incubator and a brooder is more of an intermediate/advanced skill rather than a beginner's skill. But here are some basics in case you choose to do this, or the hen goes off her brood, rejects, or cannibalizes the eggs or chicks.

Incubators

The Hova-Bator is the most popular incubator for backyard chicken raisers, for a good reason. You're going to need a brooder as well; you can get them together if you like, or just the incubator alone. [69]

Whatever incubator you choose, you need certain conditions:

- Eggs need to be kept at 99.5 degrees. One degree, higher or lower, can kill the embryos.
- Humidity is vital without the mother hen to provide it. For the first 18 days, humidity needs to be 40-50%. After that till hatching, it needs to go up to 65-70%.
- Ventilation is critical, as well. The eggshell is porous for the exchange of oxygen and carbon (of course, the eggs are *breathing*). They can suffocate, so make sure the incubator you choose has ventilation as a feature.[70]

Eggs need to be turned three times a day for 18 days. Put marks like an "X" and an "O" or "1" and "2" so that you know which side you're on. After day 18, you can leave them for the rest of the days before hatching.

The chicks will start to move and peck through their eggs. Let them hatch. It is typical for chicks to peck through and then take a rest while they adjust to breathing "outside" air instead of being in the egg. Their lungs need to strengthen, and they need a little time to do this. Don't try to help them.

Once the chick comes out of the egg, dry it off and move it to the brooder.

If you don't understand what an incubator/brooder set means, here is an example of a brooder. I like brooders with heated plates as they take much less energy than the brooders with heating lamps.

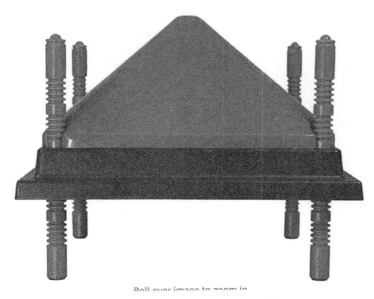

Roll over image to zoom in

Amazon http://ow.ly/JzZ950CsMHU

Breeding "Mutts"

You have to decide whether you want to keep your breeds separate or want to mix them up.

If you want to have varied breeds of hens but would love to keep the purebred breeds, then getting fertilized eggs from a hatchery is probably the way to go.

If you want to introduce a rooster to breed with the hens, then it is OK to have mutts. Their genetics will end up making them vary in egg production and behavior. I have had the luck of being able to observe a flock of 30 chickens on 17-½ acres, who are free-range. The owners have two or three roosters, depending on whether the chicks grow up to a total of 30 hens. They just decided to let them be chickens. They have separated into sub-flocks, with each rooster having 10-12 hens. The owner is actually very intentional about genetics. Their temperaments are good, they lay a lot of eggs, and they seem to have a "hybrid vigor" as far as their health goes.

It's fine to breed mutts, but make sure that you are intentional and are planning the future of your flock.

Part 6: Troubleshooting

Common health problems

Before we go into common chicken health problems, let's establish a couple of things:
1) Many of these problems can be prevented altogether by:
    - Purchasing chickens from a quality breeder,
    - ensure your chickens have all vaccinations,
    - giving your chickens proper nutrition,
    - making sure they have plenty of space,
    - regularly handling them and checking them for any injury or issues,
    - keeping them from getting bored,
    - keeping their environment clean and dry,
    - and making sure they have clean, fresh water.

    If you do not entirely prevent one of these problems, catching it early and having an extra coop for isolation, as discussed earlier, is key to saving your bird and protecting the rest of your flock.
2) Imagine if you were not familiar with dogs and thought you might want one. Then you were given a list of common health problems dogs, or cats can have. It would be very intimidating! Many of us have had dogs and cats and have never seen a list like that. Don't be intimidated by this list for chickens. Simply be aware of how to prevent any of these, how to spot them, and what to do if you experience any of these problems. Many of these ailments can be managed easily if caught early.

Diseases and Infections

A note about respiratory illnesses: Several illnesses cause respiratory symptoms in chickens. Some are common; some are not. Some are contagious; some are not. Some are easily treated or not too serious; others are fatal. If your chicken has respiratory symptoms, you can use this guide to get a sense of what the problem might be if they have other symptoms as well. But as a rule, if any of your chickens start having respiratory symptoms, call your vet. They will know what to do to properly diagnose and treat the issue; follow their guidance.

A note about diarrhea: As with respiratory illness, loose stools and diarrhea can be a symptom of a number of conditions. Is it just one chicken or several? Give the chicken(s) in question a thorough examination to look for other symptoms described in the various conditions below, and note any behavioral changes, as well. All of this is to best prepare you *when* you call your vet. You may come to a quicker diagnosis and treatment because your vet has enough information to go on. You may also save money because your vet will just suggest putting a little apple cider vinegar in their water. This guide is *not* intended as advice *instead* of calling your vet.

A note about hens not laying: As with the respiratory symptoms and diarrhea, your hens may slow down or stop laying eggs for a number of health reasons. Watch their behavior generally and examine them for any other symptoms that may point to any of the following conditions or parasites.

Fowl Pox

Fowl Pox is contagious and can be transmitted by mosquitos. If you see signs of one of your chickens with fowl pox, it is critical to isolate them and keep a close eye on the rest of the flock.

Symptoms: Hens stop laying eggs, their skin has white spots; you may find sores or ulcers in the mouth (white). Also, check their combs for scabs, spots, or sores.

There is a fowlpox vaccine; ask your vet about this for your flock. There is no treatment apart from waiting it out and keeping the chicken warm and comfortable while she heals. Chicken raisers recommend soft foods.

Infectious Bronchitis

Infectious Bronchitis is contagious. It is airborne, so isolation is important as soon as you notice the symptoms in any chicken.

Symptoms: Your hen will stop laying. It basically looks like a common cold: Coughing, sneezing, mucus/discharge from nose and eyes.

If one has it, definitely contact your vet immediately as they have a vaccine that can project the rest of your flock. Isolate the infected chicken(s) and increase their greens to fight it off.

Newcastle Disease

Newcastle Disease is another respiratory illness and is very contagious. At first, it can present like bronchitis described above but then take a more lethal turn. It is contracted through *wild* birds, so part of the preventative measures is to ensure that your shoes and clothing are not bringing in any dangers to your chicken coop. This is another reason to have shoes or boots that are specifically for your chicken area.

Symptoms: Like bronchitis, it starts with hens stopping egg-laying, having discharge like a runny nose, and also signs of finding it difficult to breathe. As the disease progresses, the neck begins to twist, and their legs and wings become paralyzed.

Even though this can be a serious and lethal disease, birds do recover. Ask your vet if to give your hen antibiotics to prevent them from contracting other infections while they're battling this disease.

You can prevent this illness with a vaccination.

Coccidiosis

This is a gut parasite that is contracted through droppings. A vaccine given to chicks before you purchase them will usually prevent this.

Symptoms: Loose droppings are the most obvious symptoms. The bird also gets lethargic and may have difficulty breathing. Sometimes the combs and wattles get swollen.

Isolate the chicken and call your vet. There is an antibiotic treatment that will kill these parasites. You'll also need to inform your vet whether your chicken can have any extra vitamins in their water when they're recovering.

Bumblefoot

This is one reason why it's important to check your chickens' feet regularly. Bumblefoot is an infected cut in the foot (usually the bottom or between the toes). Chickens scratch and run around on the ground, which may sometimes be filled with gravel or rocks. Their scaly feet are pretty tough, but they can still get cut, and just like us, cuts can get infected if not tended – especially when the feet are in poop and dirt all day.

Symptoms: You'll know it when you see it. The area of the cut swells up and is usually discolored. If not tended, the infection will progress to the leg and eventually kill the chicken.

Prevention is certainly the best cure for Bumblefoot, as surgery is the only cure if it sets in. Check for any cuts on your chicken's feet and clean them twice a day. Ask your vet for recommendations of a cleaning solution to have on hand *before* setting up your chickens. Then if a chicken has a cut foot, you are ready to tend it immediately.

Botulism

Botulism in chickens is just like humans. It comes from contaminated food. Chickens who get it are typically fed food scraps that have rotted. Meat is a common carrier, but vegetables and grains can be contaminated as well. It is not contagious, but if one chicken got it, then you must question whether other chickens were exposed to the same contaminated food source.

Symptoms: Botulism causes spasms and tremors. If you see your chicken suffering from these, you will have to contact your vet immediately. Your vet will have medication that can cure it and a plan for what to do if other chickens show symptoms. If left untended, the bird will have trouble breathing and then is paralyzed and dies a painful death.

Thrush

Thrush is another food source illness. It is a fungal infection from contaminated food or water.

Symptoms: The chicken will want to eat more than usual and will be lethargic. They begin to have a crusty vent and a thick white liquid in their crop.

Tell your vet and ask for an anti-fungal medication. It is imperative that you clean the coop and all food and water sources. Make sure that you don't have a bag of food that has gone moldy so that the problem doesn't come around again.

Always keep your chicken feed in a clean, dry container, protected from water and humidity.

Infectious Coryza

A well-named illness as it is highly infectious. It spreads through birds, water, and the chicken's environment. It is also fatal.

Symptoms: Isolate a chicken immediately if their combs and heads are swollen. There is also mucus as cold symptoms. Their eyes will eventually swell completely shut.

If you have a bird diagnosed with Infectious Coryza, it will have to be put down. There is no cure. The best way to prevent this illness is to ensure that you do not get chicks or chickens who have it or have been in an environment where they were exposed. This is another reason why a reputable breeder is essential.

Fowl Cholera

This is probably not a threat in a suburban setting as it is a zoonotic bacterial infection from wild animals. This is one reason we have recommended that you use one set of shoes for your chicken coop and run only. You can bring this home to your chickens on your shoes if you hike or are in places with wild animals.

Symptoms: Diarrhea, swollen combs, and wattles (and possibly discoloration). There is mucus from the nostrils and difficulty breathing. Some chickens may look like they are having a hard time walking.

This is a fatal disease, and there is no cure. The bird needs to be put down as, once infected, they would still be a carrier even if they recovered. There is a vaccine for it; make sure that your chickens have it.

Pullorum

Pullorum is a virus. Your chickens will get it from another bird or contaminated surfaces.

Symptoms: Sneezing and breathing problems are the major symptoms in mature birds. Chicks and younger birds will be lethargic and have a thick white paste around their vents. Sometimes birds die before it looks too serious. If this happens, and you have not contacted your vet, do so immediately to enquire about an autopsy to find out whether it was pullorum or not. Clean all surfaces, including food and water containers, clean out the coop and watch for any early signs that this fatal disease has spread.

There is no vaccine for this and no cure. If your bird is infected, it will need to be put down.

Parasites

Parasites that can multiply on your chickens include mites, lice, worms, fleas, and ticks. We are going to look at the most common ones, discuss prevention, early warning signs, and the required treatment.

Red Mites (Dermanyssus gallinae)

Red Mites are almost invisible to the naked eye, and they have a short lifespan of only about seven days, but they multiply in the 10s of thousands. They are most active from May-October and go dormant in the winter, but they **do** live through winters inside the coop for months, so don't expect cold weather to solve the problem.

Red mites can be passengers of wild birds, but in a backyard chicken coop, the most likely way they will arrive is through an introduced chicken or from clothing or a pet (e.g., a dog) that has been in a place where they have set in. You could, for example, visit a local chicken breeder and bring your dog with you, tour the chicken farm, bring mites on your clothing and bring them home to your chickens through both your clothing and your dog. Your chickens will eventually be covered with them all over, particularly around the vent.

Symptoms: Red mites feed on your chicken's blood. Your chicken will become anemic. Their combs and wattles will fade, as the infestation increases, they will stop laying eggs. Sometimes they start refusing to roost at night because the red mite hides in the cracks of the coop during the day and jumps on the birds at night.

Red mites are probably the most common problem of chicken raisers, and prevention is a lot easier than treatment.

To prevent red mites, you will need to empty, scrub, and clean your coop weekly. When you scrub it, there are two suggestions that make it easier to really get the mites out of the coop:

- Use a steamer that is used for stripping wallpaper when you clean your coop
- Use a hose with a nozzle and as much jet pressure as possible. Some chicken raisers even choose to invest in a pressure washer so that they can really get into those cracks and crevices.

If you use DE, rub it into the cracks of the coop and on the perches. Using DE as a dust bath as well as in coop cleaning has been successful for me so far; it is all I do. If you decide not to use DE, another suggestion for killing red mites in the coop is to rub a mix of paraffin and Vaseline in the cracks and corners of the coop. This suffocates them.

If you do have an infestation, then besides getting rid of the mites, consult your vet about vitamin supplements in the chicken's water to help them recover fully and more quickly.

This is the best information I have found about red mites. It helps you understand the life cycle of the mite and how best to clean, when for maximum effect.[71]

Scaly Leg Mite

Scaly leg mite is the second most common chicken mite. They are easier to deal with but may take time for the chicken to fully heal.

Symptoms: Since they are so small, it will probably take several weeks before you'll notice that there is an infestation. They live in between the scales of the chicken's legs. What you finally see is a build-up of the mite poop, fluid oozing from the legs of the chicken, and their expressions of discomfort. The mites spread from other birds, so it is likely that all of the chickens will have it if one does.

A common and simple treatment is to apply Vaseline to the legs, rubbing it into the crevices made by the scales. Mites need air, so the Vaseline will smother and kill them.

Don't forget to throw away all the bedding and to clean the coop just like you would for red mites. They don't live in the crevices of the coop in the same way as the red mites do, but they may be lurking, so you have to do a big clean when you treat your chickens.

Mites of the Feather Shafts and Bases

The red mite and scaly leg mite are the two most common mites. Other mites are:

- The Depluming Mite (this one burrows into the feather shafts) the bird's skin oozes as though it had a wound, and they may begin to pluck their feathers to try to alleviate their discomfort.
- The Northern Fowl Mite (also found at the base of the feathers, but instead of burrowing into the shaft, they are around the base of the feather)

If you see signs of either of these two mites, you will have to contact your vet for immediate treatment recommendations. The key to success is early detection. The checking of the skin around the feather shafts should be your regular weekly check-ups. Furthermore, an example of why it is so important to handle and take care of your birds at a young age is to ensure that they are tame and will not be frightened by your examinations.

Fleas

Some locations are more prone to fleas than others. Fleas can act similar to the northern fowl mite; they will be found at the base of the feathers as well as under the wings and around the vent. They can be seen by the naked eye, and like your dog or cat, your chickens will let you know that they are uncomfortable. Talk to your vet about the diagnosis of which flea it might be and follow their recommended treatment.

Ticks

Ticks will attach to your chickens as happily as they will do to you or your pets.

If you live in a place where ticks are likely, then you'll definitely have to inspect the skin of your chickens for any attached ticks.

Symptoms: If a tick is attached to a chicken, the symptoms of the chicken being the host of the tick will indicate that the crown and the wattles will be pale, as well as low energy and fewer eggs. These symptoms could be related to a number of issues, but if it is tick season, include a careful examination of the skin in your routine. If you find one attached, you can use tick tweezers as you would for yourself or another animal.

If you have a hen with one or more ticks, definitely do a big coop clean, throw out the bedding and clean all corners, crevices, nooks, and crannies.

Worms

There are a lot of different kinds of worms, and it takes a lab test to determine which one it is. Instead of living outside the body like mites, fleas, and ticks, worms live in the gut. The worm eggs are in the poop of the birds, so as the flock pecks around, there is a circular pattern to the infestation: Worm egg hatches in the gut, lays eggs, eggs are pooped out, chicken pecks in poop, gets more eggs and that's how the cycle continues.

Contact your vet for a worm test if:
- Your chickens are short of breath, especially stretching their necks and gasping for breath
- The yolks of the eggs are pale
- Chickens are eating a lot but losing weight
- Diarrhea or loose stools
- Lethargy
- Faded combs or wattles

As with most problems, prevention is the best cure. Worms are a fact of life in the spring and summer but ensuring that you do not have an ideal worm hatchery environment can do a lot better for you and your chickens. Keep both the coop and the run clean and dry. UV light kills worms, so rotating the ground or dirt will help a lot on a sunny day. Change the bedding often and keep it clean and dry as well.

Personally, I would want to get a lab test and confirm if that which I'm dealing with is a worm. After that, I would use whatever treatment the vet recommended. Some chicken raisers start with putting some apple cider vinegar or crushed garlic in the chicken's water to see if the problem can be alleviated.

If you do have a chicken with worms, it is good to treat your chickens and then move them so they aren't pecking in the same place as they were. In the meantime, you can expose that area to the sun by turning the substrate and keeping it dry.

Wounds

Wounds are serious for two reasons: 1) the risk of abscess or infection 2) hens want to pick at the blood, and they will harm or even cannibalize the wounded chicken (we will discuss this later in bullying).

Wounds can be caused by pecking or fighting, by the talons of an aerial predator such as a hawk, or an odd mishap around the yard, or roosters being too rough on your hens when mating (a common occurrence).

The regular checks on your chickens will reveal any wounds.

Stop the Bleeding and Assess the Wound

Stop the bleeding with direct pressure as you would on anyone. When you have controlled the bleeding, you can assess whether the wound is deep or surface. A surface wound can bleed a lot, so have your gauze ready, keep the pressure on and see what the problem really is. If it is a deep puncture wound or something serious like a severed wing or leg, continue to control the bleeding and call your vet. If it is a surface wound, you can continue.

Treat the Wound

It is good to have a first aid kit around for your chickens just like you might for your family.

- Nitrile gloves for handling a bleeding chicken
- Gauze
- Vetericyn Plus Poultry Spray
  - This is a non-toxic anti-microbial spray that is safe to use on your chickens.[72] Some recommend hydrogen peroxide, but since first aid certification classes teach you that peroxide is not to be applied to wounds and can be counterproductive to healing, we strongly recommend Vetericyn instead. (If the wound is around the eye, use an eyedropper or other applicator so that the spray doesn't go in the eyes.)
- Vetrap bandages[73]
- A large clean towel dedicated to chicken wrapping when necessary

Note: There are products that claim to "cover-up" the wound, therefore making it unnecessary to isolate the chicken because the wound is hidden from the other hens. I have not used these products, but I do know two chicken raisers who have used it once and never will again. I've not seen good results reported on chicken expert blogs or forums either.

Which brings us to our next point: Healing the Wound

Isolate the Chicken While the Wound Heals

Your emergency back-up chicken coop comes in handy again. Isolate your chicken and apply the Vetricyn Plus 3X/day. Care for it as you would to a wounded family member or pet. Make sure the wound is completely healed before re-introducing the chicken to the flock, and make sure you watch the behavior of the other hens closely to ensure that she is safe to be mixed back in.

Sour Crop

As we discussed in the section about feeding grit, the crop in a chicken is a pouch in the esophagus where food is stored and broken down before it moves to the stomach. The grit provided to the chicken helps break down the food since they don't have any teeth. Along with the grit, there is also a healthy amount of bacterial activity that makes the food digestible.

Sour crop occurs when the bacterial activity goes out of balance, and there is a yeast infection. The chicken's breath will be horrible (hence the name), and the food will not move to the intestines; it will be stuck.

There are directions out there for applying massage to the chicken in order to release the fluid, but this is an activity for only an advanced chicken raiser. Even then, it has a high risk if not performed correctly.

If you think your chicken has sour crop, then take her to the vet and follow their treatment and advice.

Egg Bound Hens

A hen that is egg bound has an egg trapped within her that can not be laid. This needs urgent attention and can lead to death within 24 hours.

Symptoms: You may see the hen straining with no egg coming out or going in and out of the nesting box without laying an egg. She will not be able to poop. She may waddle uncomfortably and stop eating or drinking. When you pick her up, she is likely to have a hard abdomen.

Sometimes you can deal with this yourself by placing your hen in a warm bath for ½ hour to relax the muscles and allow her to lay the egg. Gently applying some lubricant to the area after the bath can help. Having some nitrile gloves around for this purpose is useful.

If the bath doesn't work, get her to the vet. It is not recommended that you try to release the egg yourself unless you are a trained vet technician or have the training and experience to do this without harming or killing the hen.

Broodiness

A hen is "broody" when her hormones kick in to give her the maternal instinct of sitting on eggs to hatch. This is the same instinct that makes her a good mother. You have to carefully consider whether you want chickens with a strong tendency for broodiness. It is a *wonderful* trait when you want to hatch chicks, as hens that do not have a strong urge to brood will walk away from their eggs or chicks and let them die. It is *not a desired trait* for chicken raisers who do not want to hatch and raise their own chicks and just want eggs.

In the section regarding hatching chicks, we covered caring for a hen who has fertilized eggs that you intend to hatch. This next section will deal with broodiness as a problem when a hen is sitting on unfertilized eggs.

How to Tell if a Hen Has Gone Broody

A hen that has gone broody sits on her eggs in her nesting box, fluffs her feathers out, and will not move except once a day or so to poop and get some food and water. She will become much less tolerant of handling, even hostile and aggressive. She is likely to hiss, peck, and even bite when you try to collect her eggs. If you take her out of her nest box, she will run back obsessively.

The temperature of a broody hen goes up, so they can keep their eggs warm. If left, she will start to pull out her chest feathers (they do this to line the nest and heat the eggs directly).

How to Prevent or Discourage Broodiness

You may not be able to completely prevent yourself from ever having a broody hen. There are ways to discourage it, though, making it a less likely occurrence.

The first prevention is to choose breeds who are not known for going broody. Some breeds can even be a "problem" for those who want to hatch chicks because the broodiness has been bred out of them for egg-laying. We have discussed this in the breeds section of Part Two. Though there are breeds who are not particularly broody, and you want them for other reasons such as their egg color, temperament, or coloring. In that case, a little prevention can be a lot of cure.

The second discouragement of broodiness is to remove the encouraging environment where the hormonal process and maternal instincts thrive. That would be a nest with eggs. Understand the cycles and seasons of the chickens you have. Broodiness is most often an issue in the spring and summer. Some hens will lay during the winter, so they need to be watched for broodiness all year, but even these breeds will *tend* to go broody more during spring and summer.

Observe your individual chickens and know their patterns for what time of day they lay their eggs. Most chickens will lay their eggs in the morning, but breeds vary, so it is good to know their individual patterns. Starting in the early spring, make sure you collect eggs twice a day. Ideally, get the eggs right away when the hen lays it. Don't let her sit there with it; that just encourages the hormones to kick in. You will most likely know very well when your hens lay their eggs because they cackle loudly about it. Many people go to work during the day, so being there when a hen lays every egg may be impractical, but at the very least, make sure you gather the eggs in the morning before you leave and as soon as possible after your return. This discourages the hormone production for that maternal instinct and interrupts the broodiness cycle.

What to Do if You Have a Broody Hen

There are a number of things you can do if you find out that your friendly, tame hen doesn't want to leave her nesting box and is suddenly hissing and pecking at you when you collect the eggs.

- Ask yourself if you'd like to hatch some chicks. You may live in a place where roosters are not allowed, but you can purchase fertilized eggs. Or, you may want some purebred chicks from this hen, and you don't have the right rooster. Now is your chance to hatch some chicks without having the hassle of an incubator!

If you don't want to take advantage of the moment and hatch chicks, then try the following methods.

A note on handling broody hens: Wear gloves! Some people wear leather gardening gloves, but I found that I needed to use my thick welding gloves that go up to my elbows. For some hens, safety goggles might not be a bad idea. At the very least, wear gloves when you are collecting the eggs as you are likely to get pecked.

Collect the eggs at least twice a day, preferably immediately after she lays them.

- After collecting the eggs, put a package of frozen peas on the nest. This has two effects: 1) it makes the nest less comfortable and
2) if she persists despite the discomfort, it will cool her abdomen down, which will, in turn, lower her temperature and help take her out of the broodiness cycle.

- Give her a cool bath on her underside. If she doesn't sit on the frozen peas, then cool her temperature down by dipping her in a bath of cool water. Make sure to dry her off well when you're done, and obviously, don't do this if it is very cold outside.

- Block off her nesting box for the rest of the day so she can't get back in once you've removed her.

- Let her be broody. It takes 21 days for eggs to hatch and the broody cycle to complete. Another hormonal cycle kicks in to raise the chicks, and if they are not there, then the hen goes back to normal. Some chicken raisers cleverly replace the real eggs with wooden eggs and just let the hen sit until she is done.

- There are four important things to note if you decide to do this:
  1) While a hen is broody, she is obsessed with those eggs and neglects her own health. She won't bathe or get exercise or even take in enough food and water if not encouraged. If you decide to let her be broody, keep her water and food within close proximity, and make sure she is forced out a couple of times a day to eat and drink.
  2) You may need to put her in a separate box as she may get aggressive with the other hens in the nesting box. Be ready with a "Plan B" for her nesting space.
  3) She's not the only one who can get aggressive. Broody chickens are vulnerable to being bullied. When she does make her one trip a day out to get some food and water, she may be kept from accessing them and may even be pecked and wounded—another reason to separate her.
  4) Broodiness is "contagious." When one hen goes broody, others will likely follow. Letting her go through the cycle may mean that you have a henhouse full of broody hens going through their three-week cycles for weeks to months! Separating her from the flock would help with this and the other issues listed above.

- If necessary, provide a single wire cage enclosure for her until she has stopped being broody. The reason this works is that the wire on the bottom of the cage (that is off the ground) makes it uncomfortable, and she does not feel like there is a good place to nest. The air circulation will also cool down her chest and abdomen; therefore, her hormones will then shift, and she'll come out of her brood. This usually takes 3-4 days. You can check after three days to see if she's done. If she doesn't run back to her nesting box, then you know that the broodiness cycle has been broken. Make sure she has plenty of clean, fresh water and food.
  - If you think you want to use this method, plan ahead and have a cage ready for her isolation.

Soft-Shelled Eggs

A "soft-shelled" egg is an umbrella term for eggs that are soft as rubber such that you can stick your finger into it and make a dent, or very thin (yolk can be seen through the shell sometimes) or no shell at all, only a membrane.

For your young layers who are just getting started, it is quite normal for them to produce soft-shelled eggs at first. Contact your vet to find out how long this could go on before there may be an underlying cause for concern.

On the other end of the time spectrum, older hens who are nearing the end of their laying years may also start to produce soft-shelled eggs. Many people let this run its course; it's the chicken equivalent to menopause, so she will just naturally stop laying. Ask your vet about whether they recommend intervention or not.

If you have a hen in your laying years, who starts to produce soft-shelled eggs, then there may be a number of reasons.

- The first thing to consider is nutrition.

  Increase the amount of calcium in the diet; limestone or ground oyster shell are both readily available at chicken supply stores or online. They can be added to the feed or made available separately. Keeping it separate has the advantage of not having to figure out how much to add, as chickens are pretty smart about getting what they need when it is available to them.

  Are your hens getting enough sunlight? In northern regions during the winter, the days are short, and there can be inclement weather that makes the chickens want to stay inside. Try to get them out every day in order to expose them to sunlight. If you live in a northern region such as Canada, you may find that there are months where the sun is not strong enough to give a human enough Vitamin D.

  If you think that Vitamin D might be an issue for your chickens, ask your vet about Vitamin D supplements and amounts appropriate for your area.

- If you are using apple cider vinegar as a supplement or treatment for worms, then you need to be aware that too much of it can cause soft-shelled eggs. If you've assessed that both calcium and Vitamin D levels should be good, then try giving the apple cider vinegar a break, or you can use less.

- Very hot weather is a common cause of soft-shelled eggs. Chickens will tend to eat less (so less nutrition for egg production) and drink more during very hot weather spells.

- Stress is another cause of soft-shelled eggs. Protect your chickens from being harassed or chased by dogs, cats, or children and from the threat of predators. In this case, the egg is soft-shelled because it is laid before completely formed; it is laid premature. Not enough space or in-fighting can also be enough stress to cause soft-shelled eggs.

- Illness. For example, both Newcastle Disease and Infectious Bronchitis can cause soft-shelled eggs. If you have reason to believe that it is not any of the causes listed above, then watch for other signs of illness—Check-in with your vet for advice.

The Pecking Order vs. Bullying and Cannibalism

Chickens can get mean and gang up against another chicken. This is different from the normal "pecking order" establishment and can even lead to the bullied chicken being cannibalized and pecked to death. Let's look at the pecking order, followed by bullying, and things you can do to prevent and deal with bullying behavior.

The Pecking Order

Chickens are petty, and there will always be the establishment of a "pecking order" to keep order in the flock. One dominant hen will take charge; the others will follow, having status and roles all the way down the line. In a healthy flock, the pecking order will be firmly in place, but the lowest one will not be denied access to food, water, or dust baths and will not be bullied. It's just chicken status and organization.

Bullying and Cannibalism

Bullying is different. One chicken is pecked on, or feathers pulled out, as she is denied food and water and access to baths, or supplements, toys, or space in general. Sometimes other chickens can decide to reject and even kill one other chicken. It looks random to us, but there is chicken logic at play.

What Causes Bullying

A number of things can trigger a bullying response in chickens:

Illness
Chickens have an instinct to get illness out of the flock. They use murder to do this. We've discussed the importance of a second coop for isolation; contagion is not the only reason. Protecting the recovering chicken from the rest of the flock is vital.

Broodiness
Watch closely if a hen goes broody; as discussed earlier, she may be denied access to food and water when she makes her once a day jaunt. Isolation is recommended to protect her.

Over Crowding
Chickens and humans have a few things in common, and behaving badly under overcrowded conditions is one of them. Go back to the measurements you made for the chickens you acquired. Do they have ample space?

Stress
Overcrowding falls under the category of "stress", but there are other stressors that may make your chickens decide to pick on a selected victim.

Chickens are sensitive to change in routine and environment. Introducing a new member to the flock or too many at once, or even changing their feed, can trigger a stress response that results in them taking it out on one chicken to bully. Consider whether anything has changed. Did you move them into a new coop? Change is stressful for them.

Chickens are prey animals, and they know it. If they are not protected from the ground and aerial predators, they will be stressed and can act out on each other. Make sure that pets like dogs or cats do not stalk or chase them and that they have adequate protection from wild predators like hawks, raccoons, coyotes, and foxes, to name a few. Also, children need to be guided and taught to treat the chickens with kindness and not chase or scare them for "fun". It is a wonderful opportunity for a parent to nourish empathy.

Boredom

A bored chicken is frustrated and looking for something to do. Do your chickens have enough toys and variety to keep them occupied? Check out the toy section in this guide and give them lots to do.

Scarcity

Make sure you have multiple food and water stations. You need to have numerous food and water stations so that if a chicken lower on the pecking order is shooed off of one feeder, there is another behind her—the same with dust baths. Having an abundance of everything they need is a calming effect.

The same goes for toys. Don't just put up one cabbage on a rope for your flock – put up three! Watch their behavior and assess whether you need more or less. If they are fighting over the perches or swings, add some.

Roosts are very important. Ensure that there are not only the minimal square feet required but ample space for them to roost and feel like there "is always space for me." Also, know your breeds. If you have a breed of chicken that instinctively wants to roost up high and that is important to them, then, by all means, provide that space. Not doing so will produce anxiety that can come out in behavioral problems.

Give your chickens what they need. Give them enough of what they need so that they don't have anxiety about scarcity.

Wounds

Whether it be from pecking or from an accident, when there is blood, they will want to peck at it. They will peck and won't stop. Your lovely hens can suddenly turn into fierce feathered dinosaurs in your backyard as they cannibalize on one of their own.

This is one of the many reasons that handling your chickens and checking them regularly for wounds is critical. If you see any chicken that has been pecked at, have a look at the place that you saw the pecking. If there is any blood, isolate them until it is fully healed.

Ask your vet about any recommended products for the treatment and avoiding infection, but certainly, get the wounded chicken out of harm's way from the rest of the flock.

What to Do About the Bully(ies)

The first thing to do is to isolate and protect the bullied chicken *unless* it is *one* chicken that is bullying several other chickens. If this is the case, then isolate the bully.

Carefully consider whether any of the items on the list above might be the root cause of the bullying behavior (illness, wounds, broodiness, stress, overcrowding, perceived or real scarcity, boredom). Make some changes and see if the problem is eradicated. If not, you might have a particular chicken who has a dominant personality and tends to bully. Talk to your vet or a local chicken expert to see if you can find a solution for this particular chicken.

Rooster Trouble

The bullying section above is specifically geared towards hens. Roosters have their own world in this regard.

We are going to look at aggression with humans and aggression with hens separately. There is some overlap in causes and cures, but there are also important differences.

Rooster Aggression Towards Humans

The first thing to be aware of with roosters is that they will tend to be edgier during mating season. Even your docile, calm, friendly rooster may be less tolerant of being handled and act like he's trying to protect your flock. If a rooster starts to be aggressive to you or your family members, ask yourself, is it the mating season?

The second thing to be aware of is that most roosters can be guided and trained, but some will not. It's all about brain wiring. If a small child has been allowed to chase chickens, that rooster will "wire" his brain to the child being a threat. Prevent the rooster from identifying any humans as threats. Prevent harassment of your birds.

A third thing to bear in mind is that roosters (and chickens generally) should be approached with some sensitivity. Even if you are coming in to feed them, don't sneak up on them and startle them. Obviously, they will get accustomed to a daily routine, but establishing yourself as being "on their side" early on, will wire them to respond to you as a friend rather than a foe or suspect.

A fourth issue is critical to understanding how to try to train him out of his aggression: *Why* is he aggressive with you?

- Is it because it is mating season, and it is a temporary, minor edginess because he's super protective of his hens?
- Is it because something in the environment has changed or is inadequate?
- Is there enough space?
- Is his nutrition right?
  - Salt deficiency can be one key factor in rooster aggressiveness. Every three days add a tablespoon of salt to a gallon of water in the morning, then *later in the day, replace it with fresh water*.
  - Apparently, replacing chicken pellets with chicken mash can reduce aggressiveness. I believe that the key to that working is that the diet needs to have a high enough fiber content. (e.g., rolled oats, alfalfa hay, or meal). I've not seen anyone directly say this, but I asked my local chicken mentor whether the fiber addition indicated that it's actually constipation that can be a source of aggression. She thought about it for a minute and agreed. Constipation can cause humans to be grumpy; why not chickens?
- Is he bored?
- Does he have access to plenty of food, water, and dust bathing?
- Does he think he is on the top of the pecking order – including you? Is he trying to dominate you?
- Was he just "born this way"? This is where individual differentiation can play a big part. Even if you've chosen a breed known for their gentleness and have raised him and handled him yourself to try to tame him, he might just be an aggressive individual by nature. Read the story in this footnote.[74] These people know what they are doing and apply a number of the methods for training a rooster out of aggression that I have used as well. They are real "pros" and tell a story of a rooster they had hatched themselves, and in the end, even they had to get the rooster out of their flock.

After you've checked all of the factors listed above and observed changes after correcting any issues that are in your control, there are some things you can do to re-wire his brain and train him out of his misbehaving mindset.

1) Start bringing him treats by hand regularly. That way, he will start to change his association from "threat" to "friendly food source". I did this with a friend who had a rooster that went aggressive on her, and it turned him around. This might require protective gloves and caution or may not be advised at all if his aggression to you is too severe.

2) If he is trying to dominate you or your family, then you have to take charge. Each person needs to let the rooster know who is the dominant one.

   It is essential to understand that responding with dominance is **not** responding with anger or returning aggression or violence. This will only make the rooster view you as a predator and be all the more aggressive toward you. Dominance is absolutely clear, with a strong assertion that is not perceived as harmful, only dominant.

   Wear long sleeves and pants, thick work gloves (and get a large towel if you want this to help control the rooster flapping and kicking). Assertively, confidently and calmly pick him up and keep him from flapping or wrap him in the towel. Hold him firmly till he settles. It can be very helpful to slowly pet him on the back of the neck. After he has settled down, put him firmly on the ground, and make him lie down on his stomach. Do this calmly without aggression, but firmly. Put one hand on his back and use the other to press his head down. This is what another rooster who was dominant would do to him. *This is speaking his language*. After about 30-45 seconds, let him go, then be ready to do it all again if he tries it with you. Keep repeating until he stops. The rooster will not generalize that all humans are over him in the pecking order, so if there is any question of other family members, they should each do it with you as well. This video has a great demonstration of this technique around 10:40. Watching the entire video is well worth your time.[75]

3) You may have to send him to the stew pot or get another home for him. Sometimes the brain wiring just can't be shifted, and the rooster may not be a good fit for you and your flock.

Rooster Aggression Towards Hens

Sadly, roosters can be mean to hens and a threat to their health. It is important to distinguish between normal mating behavior and out of hand aggression.

The rooster will peck at the back and head of a hen to mate. She then squats down so he can mount her. This is normal and will not cause the hens lasting damage or undue stress.

Sometimes a rooster will get into a loop of aggression and abuse to his hens. You will know this is happening if you see him drawing blood or if the hens seem stressed, especially around him.

For aggression with hens, always consider the list above about environment and health (space, nutrition, boredom, etc.). Start there, make any changes and watch to see if he mellows out with the hens.

As we've discussed in the hen's bullying section, if he has drawn blood, then isolate and protect the hen and tend her wounds until she has healed.

There are a couple of protective measures that some chicken raisers take.

Do you have enough hens to have a rooster? You should have ten hens to one rooster. If he's getting frustrated with not enough hens, you'll see wounding and distress in your flock.

Pine Tar

You may see some chicken raisers suggesting pine tar. It is an old-fashioned remedy for chicken pecking, the theory being that they don't like the taste of it, so they stop the pecking. The confusing thing is that while some swear by it, others have used it and found out that it does not work at all or possibly made the pecking worse. Who to believe?

- Pine tar is anti-inflammatory, antibacterial, and anti-fungal. It is recommended for skin conditions. It has been a debate on whether it is wise to apply it to an open wound or not. Equestrians use it for their horses. Proponents say that they apply it directly to wounds, and it will protect from further infection while sucking out current pus and bacteria. [76] Ask your vet about pine tar; they might suggest another option for treating wounds. I have never used pine tar. I prefer to use Vetericyn Plus Poultry Spray and then isolate the hen so that she can rest and heal.

Chicken Saddles

A chicken saddle is a piece of cloth that is thick enough to protect from a rooster's pecking and scratching. Here is an assortment of examples,[77] and here are a number of sites that offer patterns so that you can make them yourself.[78]

Chicken saddles are great tools to have around for the specific purpose of protecting your hens from the rooster. They also come in handy in the event that one of your chickens has a wound on her back that is healing, and you want to re-introduce her to the flock.

I have been glad that I've had chicken saddles around, but there's one thing to remember. Don't just put it on and leave it. Parasites will hide under it and become a problem. You need to check it and clean it every day and check the feathers and skin underneath the area to ensure that parasites are not taking advantage of the protected environment. Also, if there are any wounds still healing up under the saddle, it is important that they have air.

We highly recommend this excellent video about rooster aggression. [79] Below is our summary of rooster aggression prevention and early intervention. (See the descriptions above for the details.)

1) **Prevention**
   - Do you have enough hens to have a rooster?
   - Check the space, the coop, food, water, toys, nutrition, and dust baths. Are they plentiful and accessible?
   - Spend time every day with your rooster – pick him up and pet him a bit. Make it normal for you to be dominant before he gets any ideas concerning it. This is not dominance in a mean way, but with a strong, quiet assertion.
   - Do you need to trim the rooster's spurs for safety? (we discuss this in maintenance)

2) Be prepared to address aggression behavior immediately if needed

Remember, you can expect the aggression to go up at least a little during mating season. Put a saddle on a chicken as soon as you think that a rooster might be too rough on her. Let the hens have some rest away from him if necessary.

Be prepared to respond right away if a rooster shows signs of trying to dominate you. Have gloves and a large towel set aside (if you choose to use one), walk out there ready to grab him if he tries to dominate or intimidate you. Follow the instructions laid out above.

A Word About Dogs

This is not a dog training guide, and each breed and an individual dog is different. I will say that I have a black Labrador Retriever and English Setter Mix who killed at least two of my neighbor's chickens, and the last one she killed, she held as her "prize".

Molly, the dog, was obsessed with sneaking past the neighbor's fence. When we finally caught her with her last "prize", we sat her down with the owner of the chicken (who is also a dog trainer), and a poor maimed chicken that my dog had almost killed that same day. I made my dog sit and held her head, my neighbor held the chicken, we brought them nose to nose, and we said, "nooooooo chickens! No!" several times. The poor chicken was not particularly happy with this activity, but it worked. I can let my dog out, and she does not go anywhere near the neighbor's fence. Not only that, when the neighbor's chickens come up to our meadow and around our house, we have always said "nooooooo chickens!". By now, Molly sees the chickens and thinks, "that's trouble", and wants to get away from them. Now she seems to think that the chickens are dangerous!

Some people have had success attaching the dead chicken to the dog collar and making their dog wear it for a day. This seems to be effective for dogs who need to learn that the chickens are not OK to kill. In our case, our dog knew very well that chickens were not grouse and not to be killed.

Don't take our word for it. Contact a dog trainer who can help. Techniques vary with breed and individuals. It is critical for your chickens to be safe and not harassed by your dog.

Conclusion

Backyard chickens are a great way to provide healthy, delicious food for your family as well as a source of fun and entertainment. They are a wonderful project to take on as a learning experience with children, as your child will learn to observe carefully and how to care for them.

This guide has discussed the important ways to set yourself up for success. Here's a quick summary:

Plan ahead. Setup your coop and think through your emergency coop before you get the chickens. Purchase your feeders, waterers, dust bath provisions beforehand, as well as their feed, grit, DE (if you are going to use it), and the storage for each of those as they can't get wet and feed will attract rodents.

Choose your breeds carefully. Start small, don't get the maximum amount of chickens that you can have in your coop. Get your routine down, and then introduce more hens gradually. Don't start with a rooster. If you want one, add him later when you've got some hen experience when your flock is settled, and you can focus on your "rooster learning curve."

Watch your chickens when you bring them home and observe how they work out their pecking order and dynamic. Be ready to put a chicken in the emergency coop if one is bleeding from pecking or one, in particular, is being a bully to the others.

Remember that plenty of clean space, abundant provisions such as food, water, dust baths, grit, and toys are necessary for the physical health of your chickens and will prevent most behavioral issues.

Enjoy the adventure and those amazing eggs!

Footnotes

[1]https://www.cbsnews.com/news/the-biggest-source-of-salmonella-outbreaks-in-u-s-may-be-in-your-backyard/

[2]https://www.almanac.com/news/home-health/chickens/raising-chickens-101-how-get-started

[3] https://modernfarmer.com/2014/07/raising-backyard-chickens-dummies/

[4] https://www.countryliving.com/life/kids-pets/a32102474/raising-chickens/

[5] https://ohlardy.com/keeping-backyard-chickens/

[6] https://www.chickens.allotment-garden.org/keeping-chickens/questions/many-eggs-hen/
https://www.almanac.com/news/home-health/chickens/raising-chickens-101-how-get-started#

[7] https://laysomeeggs.com/how-many-chickens-do-i-need-for-eggs/

[8] https://apple.news/A4CtEvznQT4OjpnqimMR0fw

[9] https://www.cdc.gov/salmonella/live-poultry-06-17/index.html

[10] Ibid.

[11] http://ow.ly/E5uD50Csw4v

[12]https://www.goodhousekeeping.com/home/gardening/a20706625/backyard-chickens/

[13] An advocacy guide written by those with experience in overturning chicken bans
https://communityfoodstrategies.org/wp-content/uploads/2020/06/Backyard-Chickens-Advocacy-Guide.pdf

Sites that are very useful and can also offer connection to get support:
https://communityfoodstrategies.org/2020/06/08/backyard-chickens-in-the-city/#section-one
https://the-chicken-chick.com/legalizing-backyard-chickens-from/
https://www.peteandgerrys.com/blog/local-laws-for-raising-backyard-chickens

A quick search will show multiple articles about people advocating for the legalization of chicken raising in their municipalities:
https://dailygazette.com/article/2019/08/26/advocates-hatch-campaign-to-legalize-backyard-chickens
https://www.dailyrepublic.com/all-dr-news/solano-news/vacaville/vacaville-residents-advocate-for-backyard-chickens/

This one is a great example of a presentation to the city:
https://d3n8a8pro7vhmx.cloudfront.net/gardenworksproject/pages/31/attachments/original/152099745
5/Advocating_for_DuPage_County-
Wide_Allowance_of_Backyard_Poultry_web_text_%281%29.docx.pdf?1520997455

[14] https://modernfarmer.com/2014/07/raising-backyard-chickens-dummies/

[15] https://morningchores.com/chicken-coop-plans/#free-plans
https://www.construct101.com/5-free-chicken-coop-plans/
https://steamykitchen.com/20640-the-palace-chicken-coop.html
https://www.buildeazy.com/chicken-coop/5/#The-front-elevation-plan
https://www.countryliving.com/diy-crafts/g2452/diy-chicken-coops/?slide=6

[16] https://www.consumerwatch.com/construction/pressure-treated-lumber/

[17] https://www.consumerwatch.com/construction/pressure-treated-lumber/

[18] https://www.naturalhandyman.com/iip/infxtra/infpre.html

[19] https://www.google.com/search?q=recycle+playhouse+as+a+chicken+coop&sxsrf=ALeKk02bs6p7_p0B
mT2MrLKX_xoHEalyRA:1600553357797&source=lnms&tbm=isch&sa=X&ved=2ahUKEwjTntfynfbrAhVjLH0
KHcDqAHgQ_AUoAXoECBAQAw&biw=1902&bih=937

[20] https://www.tractorsupply.com/tsc/product/overez-chicken-feeder-oezcf

https://www.wayfair.com/pet/pdp/ware-manufacturing-chicken-feeder-wfg1284.html
https://www.wayfair.com/pet/pdp/archie-oscar-gracie-precious-poultry-feeder-and-waterers-
aosc1655.html
https://www.wayfair.com/pet/pdp/miller-mfg-hanging-poultry-feeder-with-pan-mfg1130.html

This feeder includes a "rodent-proof" feature
https://www.wayfair.com/pet/pdp/archie-oscar-greta-poultry-feeders-and-waterers-aosc1656.html

https://www.hayneedle.com/product/littlegiantfarmagmillermanufacturing12quarthookoverfeeder.cfm

[21] Waterers

https://www.amazon.com/RentACoop-Gallon-Automatic-Chicken-
Waterer/dp/B07G7C572D/ref=sr_1_12?dchild=1&hvadid=78134172272081&hvbmt=bb&hvdev=c&hvqmt
=b&keywords=chicken+waterer+heater+base&qid=1597258446&sr=8-12&tag=mh0b-20

https://www.wayfair.com/pet/pdp/animal-house-waterer-3-oz-vvf10008.html

https://www.tractorsupply.com/tsc/product/farm-tuff-top-fill-poultry-and-game-bird-waterer-5-gal https://www.tractorsupply.com/tsc/product/rugged-ranch-high-end-hen-poultry-waterer

Heating Units for Waterers
https://www.amazon.com/Farm-Innovators-Chicken-Waterer-Deicer/dp/B01I57ZFEU/ref=sr_1_5?dchild=1&hvadid=78134172272081&hvbmt=bb&hvdev=c&hvqmt=b&keywords=chicken+waterer+heater+base&qid=1597258446&sr=8-5&tag=mh0b-20

https://www.amazon.com/Farm-Innovators-HP-125-Poultry-125-Watt/dp/B000HHQDPM/ref=sr_1_3?dchild=1&hvadid=78134172272081&hvbmt=bb&hvdev=c&hvqmt=b&keywords=chicken+waterer+heater+base&qid=1597258446&sr=8-3&tag=mh0b-20

https://www.amazon.com/Brower-AEB-Automatic-Electric-Heater/dp/B003OAPG8Y/ref=sr_1_4?dchild=1&hvadid=78134172272081&hvbmt=bb&hvdev=c&hvqmt=b&keywords=chicken+waterer+heater+base&qid=1597258446&sr=8-4&tag=mh0b-20

Waterers with integral heating units:

https://www.amazon.com/Farm-Innovators-HPF-100-All-Seasons-Fountain/dp/B001CCSJCQ/ref=sr_1_6?dchild=1&hvadid=78134172272081&hvbmt=bb&hvdev=c&hvqmt=b&keywords=chicken+waterer+heater+base&qid=1597263018&sr=8-6&tag=mh0b-20

https://www.amazon.com/Rite-Farm-Products-Poultry-Chicken/dp/B01AYHWVAU/ref=sr_1_7?dchild=1&hvadid=78134172272081&hvbmt=bb&hvdev=c&hvqmt=b&keywords=chicken+waterer+heater+base&qid=1597258446&sr=8-7&tag=mh0b-20

[22] https://homesteadwishing.com/chicken-pecking-block-recipe/
https://www.onehundreddollarsamonth.com/diy-homemade-flock-blocks-for-chickens/
https://www.thehappychickencoop.com/homemade-diy-flock-block/

[23] https://www.youtube.com/watch?v=ReC85qqSpHA

[24] https://www.youtube.com/watch?v=xSgFDxMKoH0

[25] https://www.mypetchicken.com/catalog/Chicken-Gifts-and-D-eacute-cor/Chicken-Swing-p1548.aspx

[26] https://www.backyardchickens.com/articles/diy-chicken-swing-for-the-brooder.74466/

[27] https://www.backyardchickencoops.com.au/blogs/learning-centre/5-things-add-chicken-dust-bath

[28] https://www.wideopenpets.com/dust-bathing-from-a-to-z-for-backyard-chickens/

[29] https://www.hobbyfarms.com/11-options-for-coop-run-bedding/

[30]https://lucernefarms.com/koop-cle

https://talmagefarm.com/catalog/product/163823/koop-clean-chicken-bedding-2-4-cu-ft#:~:text='KOOP%20CLEAN'%20is%20an%20all,reduce%20moisture%20in%20your%20coop.

[31]https://www.tractorsupply.com/tsc/product/standlee-premium-western-forage-certified-chopped-straw-25-lb?cm_vc=-10005&cm_mmc=Affiliates-_-Pepperjam-_-Generic-_-Offer&affiliate_id=21181&click_id=3238448835&utm_source=pepperjam&utm_medium=affiliate&clickId=3238448835

[32]https://www.amazon.com/MagJo-Excelsior-Shaving-Nesting-Liners/dp/B076ZSJBWD/ref=sr_1_11?dchild=1&keywords=Koop+Clean&qid=1597353030&sr=8-11

[33] https://www.tractorsupply.com/tsc/product/tractor-supply-co-flake-premium-pine-shavings-covers-8-cu-ft?cm_vc=IOPDP2

https://www.amazon.com/RentACoop-Hemp-Bedding-1-Pack/dp/B075SP4X2W/ref=sr_1_9?dchild=1&keywords=Koop+Clean&qid=1597353030&sr=8-9

[34] https://www.raising-happy-chickens.com/diatomaceous-earth.html

[35]This article is a well-balanced look at the data for the uses of diatomaceous earth
https://www.raising-happy-chickens.com/diatomaceous-earth.html

Here are more resources for your own Diatomaceous Earth Research
https://draxe.com/nutrition/diatomaceous-earth/
https://www.raising-happy-chickens.com/diatomaceous-earth.html#sources
https://www.raising-happy-chickens.com/diatomaceous-earth.html
Occupational Safety and Health Administration: Crystalline Silica Exposure - Health Hazard Information. Pub. 2002.
Bunch, T. R.; Bond, C.; Buhl, K.; Stone, D.: Diatomaceous Earth. Pub. National Pesticide Information Center, Oregon State University, 2013.
Bennet, D., et al: Effect of diatomaceous earth on parasite load, egg production and egg quality of free-range organic laying hens. Pub. Journal of Poultry Science, 2011.
https://www.backyardchickencoops.com.au/blogs/learning-centre/diatomaceous-earth-a-natural-remedy-for-troubled-chickens

[37] https://www.roysfarm.com/rhode-island-red-chicken/

[38] https://www.thehappychickencoop.com/leghorn-chicken/

[39] https://www.youtube.com/watch?v=copoFFyD0gs

[40] https://www.goodhousekeeping.com/home/gardening/a20706625/backyard-chickens/

[41] https://extension.uga.edu/publications/detail.html?number=C954
https://extension.uga.edu/publications/detail.html?number=C954

[42] https://www.acehardware.com/departments/home-and-decor/trash-and-recycling/garbage-cans-and-recycling-bins/71496?store=16809&gclid=Cj0KCQjwhvf6BRCkARIsAGl1GGgwD-32VH2dG6Fit7KB-VOG66Ez9OXmVcL0CwLSYxl9pwXgKLCA3JQaAr0sEALw_wcB&gclsrc=aw.ds

[43] https://www.amazon.com/Grubblies-Mealworms-USA-Grown-Nutritious-Oven-Dried/dp/B06WW6R3RD/ref=as_li_ss_tl?keywords=chicken+treats&qid=1550202297&s=gateway&sr=8-1-spons&psc=1&linkCode=sl1&tag=wop_general-20&linkId=7c4dbb296fbaf74769422c72d6b407c9&language=en_US

[44] https://cast.desu.edu/sites/cast/files/document/16/pumpkin_seeds-worms_djo.pdf

[45] https://the-chicken-chick.com/?s=How+to+clean+eggs

[46] https://www.thehappychickencoop.com/washing-eggs/

[47] https://www.mypetchicken.com/catalog/Chicken-Supplies/Egg-Cleanser-Concentrate-16-oz-p768.aspx

[48] https://backyardpoultry.iamcountryside.com/eggs-meat/do-eggs-need-to-be-refrigerated-and-washed/

[49] https://www.mypetchicken.com/catalog/Chicken-Supplies/Egg-Cleanser-Concentrate-16-oz-p768.aspx

[50] We recommend you start here. This article offers scientifically based and common sense solutions.
https://backyardpoultry.iamcountryside.com/eggs-meat/how-to-wash-fresh-eggs-its-safer-not-to/
Two other articles by credible chicken experts with varying recommendations are here:
https://the-chicken-chick.com/?s=How+to+clean+eggs
https://www.thehappychickencoop.com/washing-eggs/

[51] https://backyardpoultry.iamcountryside.com/eggs-meat/do-eggs-need-to-be-refrigerated-and-washed/

[52] https://www.amazon.com/gp/product/B006MXCM6E/ref=as_li_tf_tl?ie=UTF8&camp=1789&creative=9325&creativeASIN=B006MXCM6E&linkCode=as2&tag=thechichi-20

[53] We highly recommend this resource for trimming beaks, claws and spurs
https://backyardpoultry.iamcountryside.com/feed-health/how-to-trim-chicken-beaks-claws-and-spurs/#:~:text=Claw%20Trimming&text=Chickens%20evolved%20in%20an%20environment,the%20bird's%20comfort%20or%20safety.

[54] https://www.backyardchickencoops.com.au/blogs/learning-centre/introducing-new-chickens-to-your-flock

[55] https://www.amazon.com/s?k=chicken+waterer+heated&ref=nb_sb_noss

[56] https://farmingmybackyard.com/winter-chicken-care/

[57] https://the-chicken-chick.com/caring-for-broody-hens-in-extreme-cold/

[58] https://the-chicken-chick.com/caring-for-broody-hens-in-extreme-cold/

[59] https://www.purinamills.com/chicken-feed/education/detail/three-tips-to-help-molting-chickens

[60] https://www.amazon.com/Grubblies-USA-Grown-Mealworms-Nutritious-Oven-Dried/dp/B07B43Q4SS/ref=sr_1_7?dchild=1&keywords=grubblies+chicken+treats&qid=1600279648&sr=8-7

[61] https://www.airgundepot.com/pellet-pistols.html

[62] https://www.amazon.com/21833-Stainless-Processing-Integrated-Irrigation/dp/B01BI5D0MK

[63] https://www.youtube.com/watch?v=FzCMgmyBeZI

[64] Go to step 07 of this page to continue processing the chicken after plucking.
https://www.thespruce.com/slaughter-chickens-for-meat-3016856

[65] https://www.thespruce.com/slaughter-chickens-for-meat-3016856

[66] https://the-chicken-chick.com/caring-for-broody-hens-facilitating-egg/

[67] https://www.mannapro.com/homestead/new-chicks-in-the-flock

[68] https://ifacountrystores.com/2020/chickens/introduce-new-chicks-to-flock/

[69] Hova-Bator and Brooder Combo
https://www.amazon.com/HovaBator-Genesis-Ultimate-Incubator-Brooder/dp/B004XNVE96/ref=sr_1_4_sspa?crid=1U8MBNN56N66U&dchild=1&keywords=hova-bator+incubator&qid=1600550634&sprefix=hova-bator%2Caps%2C283&sr=8-4-spons&psc=1&spLa=ZW5jcnlwdGVkUXVhbGlmaWVyPUEyOFZaVjVGNElNQVpJJmVuY3J5cHRlZElkPUEwNzg0MzY4MUtHRUdFRFJBNkdCMiZlbmNyeXB0ZWRBZElkPUEwMDAyOTg4MzdGTVc3NTRXTFNVRyZ3aWRnZXROYW1lPXNwX2F0ZiZhY3Rpb249Y2xpY2tSZWRpcmVjdCZkb05vdExvZ0NsaWNrPXRydWU=
Hova-Bator stand alone
https://www.amazon.com/HovaBator-Advanced-Incubator-Combo-Kit/dp/B004XNVCS4/ref=sr_1_11?crid=1U8MBNN56N66U&dchild=1&keywords=hova-bator+incubator&qid=1600550827&sprefix=hova-bator%2Caps%2C283&sr=8-11

[70] https://modernfarmer.com/2015/04/how-to-incubate-chicken-eggs/
https://www.purinamills.com/chicken-feed/education/detail/hatching-eggs-at-home-a-21-day-guide-for-baby-chicks

[71] https://poultrykeeper.com/external-problems/red-mite/

[72] https://vetericyn.com/product/vetericyn-plus-poultry-care/

[73] https://www.amazon.com/3M-Vetrap-Bandage-Tape-Yard/dp/B002EAQ0OY

[74] https://www.hobbyfarms.com/how-to-deal-with-aggressive-roosters-3/

[75] https://www.youtube.com/watch?v=l91CcqwnNgg&app=desktop

[76] https://www.horseforum.com/horse-health/anyone-apply-pine-tar-wounds-injuries-153517/

[77] https://www.amazon.com/s?k=chicken+saddle&hvadid=77859218470152&hvbmt=be&hvdev=c&hvqmt=e&tag=mh0b-20&ref=pd_sl_3yr5q8slfz_e

[78] https://www.motherearthnews.com/diy/how-to-make-a-chicken-saddle-zbcz1604
https://www.walkerland.ca/amp/chicken-saddle/

[79] https://www.youtube.com/watch?v=l91CcqwnNgg&app=desktop

Printed in Great Britain
by Amazon